PENGUIN BOOKS

THE VINEYARD

P9-AFI-601

Louisa Thomas Hargrave is a former owner of Hargrave Vineyard, established in 1973 in Cutchogue, New York, near her current home.

Praise for *The Vineyard*

"You don't look for covered wagon stories as close to home as this, but everyone who plants a new vineyard is a pioneer, and the Hargraves of North Fork were backing a pretty far-out hunch: the claret of the Hamptons. It is a tale of true grit, psychological and physical. It is also the true story of wine as only a learner can tell it. I have never seen the how and why of wine so firmly drawn, nor so closely linked to the other facts of life."

—Hugh Johnson, author of *The World Atlas of Wine*

"Good women and good writing ripen like fine wine and Louisa Thomas Hargrave combines the three in this tender and truthful memoir of her marriage and her wins and losses in both love and business. At the end, one can sigh—'that was a very good book.'"

—Laura Shaine Cunningham, author of *Sleeping Arrangements* and *A Place in the Country*

"Alex and Louisa Hargrave did more than start a winery on Long Island, they created a wine region. With warmth and spirit, Louisa recounts their thirty-year saga from happy early days to a poignant end when she starts a new life on her own. A wine story? Yes, but more than that."

—Frank Prial

"Louisa Hargrave's story of growing grapes and making wine successfully on Long Island's North Fork is about a dream brought to reality through perseverance, hard work and dedication. This book is must reading for anyone who thinks the wine business is easy, and provides original insights for the wine connoisseur. In her engaging personal voice, Louisa Hargrave describes the challenge, the science and the art of making wine in an untested region."

—Harriet Lembeck, author of *Grossman's Guide to Wines, Beers and Spirits,* Seventh revised edition

"The hundreds of people, maybe thousands, who know Louisa Hargrave appreciate that she is a straightforward, realistic and good-natured lady for whom nonstop work seems a necessity. All of this comes through in the domestic details in *The Vineyard*. She is engrossing in telling about wrestling with life's problems, especially big ones. . . . The book is chockablock with technical information, but only in a way that advances the story. Planting, grafting, pruning, picking, winemaking methods, yeasts, blending—these matters are presented in an ungeeky, down-to-earth style."

—Howard Goldberg, *The New York Times*

"If you could make great wine scientifically, we'd be drinking Coca-Cola Red. But wine is the rare product that stands or falls on intangibles: love, mindfulness, wit. Louisa's book would be riveting if it only told us how she and her husband Alec mastered the science; what makes it an instant classic is her record of their more soulful quest. It was all I could do not to open a bottle, the better to savor every word."   —Jesse Kornbluth, Bookreporter.com

"Youth and naivete—and, on this case, quite a few grapes can be the perfect ingredients for success."—Heather Von Tesoriero, *Time*

"Good writing about wine is extraordinarily rare, but Louisa Hargrave pulls it off, instantly putting herself in the same company as A. J. Liebling and M. F. K. Fisher."   —Tony Hendra

"Part memoir, part history . . . the stories in her book are shaped by the hand-hoe as well as the pen, colored by long workdays, unexpected troubles with regulatory agencies, the politics of wine in New York State and the emotions of both creating and letting go."

—Peter Gianotti, *Newsday*

"At heart, Hargrave's book is a coming-of-age story that happens to take place in a vineyard. . . . Often humorous, it is a poignant tale of discovery, loss and self-acceptance. Long Island is lucky to have her."   —*Wine Spectator*

"In her candid, bittersweet memoir, Hargrave tells how they over-came hurricanes, destructive birds, diseased plants, problems with regulatory agencies, and jealous wine experts who wished them ill to achieve their goal of growing *Vitis vinifera* and producing award-winning wines. . . . As she looks back on years of joy as well as hard work, Hargrave presents a colorful picture of life at the vineyard."
—*Publishers Weekly*

"The story catches the pioneer feel of the venture: plain, fraught, moments when Hargrave thinks she's the luckiest person in the world, and then the opposing winds—personal meteorological, economic—that buffet all settlers to new country."
—*Kirkus Reviews*

"*The Vineyard* is more than an inside view of what it took to create a pioneering vineyard. It is a compelling, deeply personal story of a life and a marriage and a search for happiness with which we can all identify." —Michael Braverman, *The East Hampton Star*

"This in-depth look into the inner workings of the wine world is also a lyrical and poignant personal story."
—*Les Dames d'Escoffier International Journal*

"If you have ever drunk a glass or two or a bottle or two of Long Island wine, if you enjoyed Frances Mayes' *Under the Tuscan Sun* or Peter Mayles' stories about trying to settle in Provence, then this book should be a must read for you. Even if you don't come under these categories but enjoy a well crafted, heartwarming story of modern day pioneers, then, again, put this book on your must-read list." —Roy Bradbrook, LongIslandWineCountry.com

Louisa Thomas Hargrave

The Vineyard

A Memoir

PENGUIN BOOKS

PENGUIN BOOKS

Published by the Penguin Group

Penguin Group (USA) Inc., 375 Hudson Street, New York, New York 10014, U.S.A.

Penguin Books Ltd, 80 Strand, London WC2R 0RL, England

Penguin Books Australia Ltd, 250 Camberwell Road,
Camberwell, Victoria 3124, Australia

Penguin Books Canada Ltd, 10 Alcorn Avenue, Toronto, Ontario, Canada M4V 3B2

Penguin Books India (P) Ltd, 11 Community Centre,
Panchsheel Park, New Delhi – 110 017, India

Penguin Books (N.Z.) Ltd, Cnr Rosedale and Airborne Roads,
Albany, Auckland, New Zealand

Penguin Books (South Africa) (Pty) Ltd, 24 Sturdee Avenue, Rosebank,
Johannesburg 2196, South Africa

Penguin Books Ltd, Registered Offices: 80 Strand, London WC2R 0RL, England

First published in the United States of America by Viking Penguin,
a member of Penguin Group (USA) Inc., 2003
Published in Penguin Books 2004

1   3   5   7   9   10   8   6   4   2

Copyright © Louisa Thomas Hargrave, 2003
All rights reserved

THE LIBRARY OF CONGRESS HAS CATALOGED
THE HARDCOVER EDITION AS FOLLOWS:
Hargrave, Louisa Thomas.
The vineyard : the pleasures and perils of creating an American
family winery / Louisa Thomas Hargrave.
p.   cm.
ISBN 0-670-03221-2 (h.c.)
ISBN 0 14 20.0431 6 (pbk.)
1. Hargrave, Louisa Thomas.   2. Viticulturists—New York—Long Island—
Biography.   3. Vintners—New York—Long Island—Biography.   I. Title.
SB387.682.H37A3 2003
634.8'092—dc21
[B]   2002044901

Printed in the United States of America
Set in Garamond Three, with Sackers Italian Script
Designed by Carla Bolte
Drawing by Rob White for Hargrave Vineyard Label, 1976

For Anne and Zander

"The Jother"

Come, ye thankful people, come;
Raise the song of harvest home.
All is safely gathered in
'ere the winter storms begin.

—Traditional Thanksgiving hymn

Acknowledgments

FOR MANY YEARS, I WANTED TO TELL THE STORY OF WHAT IT was like to plant a vineyard in a new region, and to raise a family in the midst of it. It was only after the vineyard was no longer mine that I had an opportunity to do it.

I would like to thank my brother, Evan Thomas, for encouraging me to write in the first place. For selflessly offering me their homes for refuge, their hearts for comfort, and their laughter for salvation during the time that I was writing, I especially wish to thank my aunt Frances Gates and my friends Nina Moriarty, Ruth Robins, Mary Lindsay, and Connie Pim.

I am also grateful to many other people who helped me while I was writing this book, including Gail Wickham; Eric Bressler; Paulette Satur; Eberhard Muller; Beverly Barbour-Soules; Lisa van de Water; Lauren Fortmiller; Peter Sichel; Charlotte Hanson; Sarah Hoskins; Mare Lindsay; my sister, Wendy Zara; and my sister-in-law, Oscie Thomas.

As I worked to define and refine the book, Alison Clarke illuminated my path, while Paul de Angelis gave me the tools to organize my story and helped me past my mental blocks. Amanda Urban at ICM helped steer me in the right direction and led me to Denise Shannon, my agent, who stayed with the project until it found its proper home with Kathryn Court at Viking. Kathryn and Viking editor Sarah Manges guided my revisions in the most gentle and diplomatic way, for which I am eternally grateful.

Without the love, support, and astute comments of my mother, Anne Thomas, my son, Zander, and my daughter, Anne, I would never have dared to begin this project, much less complete it.

This book is faithful to my memory of events as they happened, but I have changed some names for the usual reasons.

Contents

Prologue   1

1   The Homestead   5

2   Farm Labor   30

3   Restless Beginnings   50

4   Westward Ho! and Back   72

5   Mother Nature   89

6   Fruition   113

7   Survival of the Fittest   133

8   Pet Fever   142

9   Growing Up   152

10   Taking Care of Business   165

11   The World at Our Door   175

12   Arrows in the Back   195

13   Just Plain Folks   208

14   Scalped   229

15   Moving On   239

Epilogue   247

Notes   253

The Vineyard

Prologue

I WAS STANDING IN THE KITCHEN OF MY LONG ISLAND FARM-house, looking out the window, when the phone rang. The caller was a woman who had read about our new wine venture in the paper. "You're a pioneer, aren't you?" she asked. "I've been called that," I replied. "Well, come quickly," she said. "Pioneers know how to deliver babies, and my neighbor is giving birth!"

I laughed at the thought that I might be a midwife. And while I didn't deliver that baby, it was true; my husband and I were pioneers, but pioneers with a twist. We were college-educated suburbanites who had fallen in love with each other and with wine. We decided to make a life together, growing grapes and making wine, and we spent half a year exploring the wine regions of the West Coast before deciding to return to our native East Coast, where we planted wine grapes on a Long Island potato farm. Everyone said that wine grapes wouldn't grow there, but I became the midwife for a whole new wine region.

At first, we thought of ourselves more as poets than as pioneers, but nearly thirty years of growing tender grapevines on Long Island has proved that it was our pioneering enthusiasm to get the job done and to challenge the legacy of failure for this type of crop that made a success of this new viticultural area.

The North Fork of Long Island became a farming frontier with three thousand acres of vines, in a place where rampant suburban development had been a foregone conclusion. In a letter, *Gourmet* magazine's Gerald Asher wondered, "Was it Talleyrand or Brillat-Savarin who said the invention of a new dish contributed more to human happiness than the discovery of a new star? What would either of them have said to the idea of a whole new wine region?"

Like other pioneers, we had to overcome natural disasters, pestilence, varmints, and folks who wanted to shoot us down. As the rhythms of nature gained ascendancy over us—two people who had never before been attuned to its tempests and harmonies—the joys and challenges of family life were complicated by the demands of the business. For me, to be so pregnant that I had to hold my breath while bending over to plant a vine; to stand, arms outstretched, in the face of a hurricane just to feel its force; to stomp grapes in a vat when the press broke down— these were welcome challenges, a part of the frontier life. But I wasn't prepared for the arrows that were shot at our backs. How could I know that the avuncular professor who came to advise us hoped we'd fail, that the certified plants I tended so carefully were riddled with virus, or that the Bangladeshi woman I had hired would be visited once a week by a slave master who beat her for her wages?

We had to learn all the basics from scratch, from planting to bottling. We also learned about the ways of the wine world, with its princes and professors all vying for a claim to fame, willing to destroy one another for it. I discovered that winemaking is both a spiritual and a scientific endeavor. I learned to let the vine and the wine speak for themselves, instead of imposing my ideas of how they should be crafted.

While I often approached wine in an analytic way, I wasn't immune to surprises. Once, I went into the chilly cellar, where the wine lay in barrels, with a sample of a wine I hadn't tasted all winter. It was a Cabernet that hadn't shown any particular promise. When I put my nose into the glass and inhaled an aroma that exactly matched what I had always hoped for but never really believed was possible to create, I burst into tears.

Usually, I kept those emotions under wraps. I liked the relaxation wine brought, but I felt I couldn't drink too much or I might let my guard down. There was too much at stake, too much work to do, for me to let myself go. Maybe a little drunkenness would have served me well.

The first day I spent farming, following the tractor with a grape planter, was a model for the next thirty years. Under the open sky, work fell into a rhythm that was at first boring and then consoling. My mind could go anywhere because the work didn't require much new thinking. I could feel the power of my body increasing as I worked every muscle in my 110-pound frame.

At the end of that first day, I made a pact with the last plant I put into the ground. It was at the end of a row, near the woods to the east. I promised the plant that as long as it lived, I would stay on the farm and take care of it. In the end I couldn't keep my promise. When I left the farm I made a last stop with apologies to that vine, recently unburdened of its fruit, its leaves yellow and half fallen to the ground, but very much alive.

Despite my broken promise, I am thrilled that the vine has lived. It vindicates all my love, all my effort and attention. I am thankful for those years of marriage and family and farming. Even now, the vine shows the promise of renewal, and its wine offers the example of change for the better.

I

The Homestead

ON THE FIRST DAY OF JUNE IN 1973, I DROVE MYSELF FROM Rochester, New York, to our new farm in Cutchogue, eighty-five miles from Manhattan on the North Fork of Long Island. My husband, Alex, had already moved down there to start planting at the beginning of May, but I had stayed behind to finish the courses in calculus and chemistry that I was taking at the University of Rochester. We had never grown grapes or made wine before, and one of us needed to learn some science. In the four years of college and one year of graduate school I had recently completed, I had avoided studying science. I had passed the required "science for poets" course by memorizing questions and answers from past exams, not by understanding the material. History and languages were more my thing, but now that we were planting a vineyard, I wanted to understand what lay behind the mystery of wine. While I studied theories of solutions and exothermic reactions, Alex studied the more practical issues: What kind of farm equipment did we need? Where could we find the right grape varieties? How should we plant the vines, and what would come next? He was the one, we agreed, who would figure things out and get the ball rolling. I was twenty-five and Alex was twenty-seven. With no farm experience and little life experience, we really didn't think the vines would need much attention.

Before we bought the farm in Cutchogue, neither one of us had grown so much as a vegetable garden. That is, unless you count the tiny garden that had been my family's project when I was about eight years old. My father, who worked in Manhattan as a book editor and knew nothing about plants, had gotten a bee in his bonnet about wanting a vegetable garden—something that had never interested him before. We lived in the woods where the soil was poor and there was no sunlight, but still my father, mother, sister, and I hacked away at a six-foot-square plot of clay while my baby brother threw clods of grass around. Every day we checked the garden to see if anything had grown, and I remember jumping up and down with my sister, Wendy, when some tiny radish leaves sprouted. After we harvested the radishes, the garden got weedy and we gave up on it. I felt cheated by the radishes; they looked so inviting in their fabulous red skins, but they tasted unbearably bitter. We never planted another vegetable. I didn't learn much of anything about plants until after I was married and I was preparing to grow acres of grapevines.

Before making the five-hundred-mile drive to Long Island I should have been entirely focused on becoming a vintner. I certainly wasn't paying much attention to my studies. In fact, I was unable to focus on much of anything. I was numb, except for my jaw, where every ounce of tension in my body was stored. One night, about a week before moving, I gave in to the physical pain and cried myself to sleep. I never told anyone that I was anything but happily excited about my future as a grape grower. For me it was "all systems go" once the decision was made to plant grapes. Never once did Alex and I have any sort of conversation that would have permitted the shadow of doubt to fall over our plans.

The idea of the vineyard at that point was still a fantasy

whose only tangible basis in reality lay in the ten thousand rooted, grafted vines we had bought. On the first of May, we had packed our vines into Alex's brother Charlie's VW Beetle convertible for the trip to Long Island. Alex had enlisted Charlie and their sister Meg, both fresh out of high school and with no other immediate plans, to accompany him with this load, and to help with the planting until I could get down there. There was Charlie, his long hair and shaggy beard helping to hide his shyness, which was belied by the deft way he loaded the car. Meg was equally capable, but her quick, sardonic laugh was anything but shy.

The vines were bundled in plastic bags to keep their roots moist. Peeking into one of the bags before the others took off for Long Island, I noticed the delicacy of the vines. Some buds, pushing up prematurely in the darkness of the bag, were a fleshy white, like tangled bean sprouts. Before we could plant the vines, we would have to chop their roots with a hatchet, so that they would fit into the plowed furrow. That's what we had learned from the University of California textbook called *General Viticulture* that we had bought a few months before. It was a book we were relying on to teach us everything about grapes. The authors of that volume were pretty explicit about what to do with the plants, but they neglected to address the needs of the planters. What about our own roots, which were chopped, too, as we moved from everything that was familiar in our bookish suburban lives to the pastoral greenbelt of the North Fork of Long Island? That the farmland looked nothing like a wilderness was a gentle deception.

Making the drive from Rochester to Cutchogue, impatient with the time it took to get there, I was oblivious to how fundamentally my life was about to change. I might as well have been sleepwalking through time and space. When I left Rochester on that June day, I was a student; when I arrived in Cutchogue, I became a farmer instantly. Before the move, the time spent planning the vineyard had been no different from preparing for a science or a history exam, something I was familiar with. How many years of history exams had I taken? Do a bunch of reading, be a little intellectually excited and a little bored, come into a roomful of familiar people all rustling papers, and write like crazy knowing just what the professor wants. I had always done well at that. But what good would those skills do me now? In preparing for this new life, I had only the vaguest sense that there would be a massive disconnect between my old life and my new one. Whenever feelings of anxiety arose I would tell myself, "Don't have a kitten." The irony of it is, that's exactly what I did—I adopted a kitten, a real one.

The day before I left Rochester, an acquaintance had called me, urging me to adopt a six-week-old kitten from her Siamese cat's new litter. Her cat was a prima donna who had lost her virtue to a neighborhood rapscallion, a black tomcat. I sorted through the kittens and noticed one, a black female with a white patch under her chin, staring at me. That was the one I adopted. When it came to naming her, I heard in my mind my old Latin teacher saying, "I don't give one iota what you think," and I called the kitten Iota.

I was preoccupied with Iota as I drove myself from Rochester to Cutchogue, a nine-hour trip in a car with no radio, just a squalling kitten. I hadn't told Alex about Iota—maybe I hadn't wanted to run the risk of having him say that it was too soon to be adopting a pet—so I was worried about how he might react. When I saw the sign for Cutchogue on the highway, I realized that I didn't have a clear idea of what the farm we had just bought looked like. I'd only seen it once, in the pouring rain, five months previously.

It was late in the afternoon as I entered the farm. I took a deep breath to fortify myself as I made the turn south down a pitted dirt road. First I passed the two ranch houses of the Zuhoskis, the family who had set aside a few plots for themselves when they sold the farm to a developer, who had then sold it to us. Their houses were new, with only a few locust trees dividing their carved-up lots. I gave no thought to what these neighbors would be like until later in the week, when young Jeannie Zuhoski came walking down the road, carrying a pie to welcome us. Had I arrived a few years later, I would have had to contend with some new neighbors who hated the dust that blew into their windows every time we drove past them, and who came out screaming and swearing when they saw us. This first time, there was silence.

There were potatoes planted on the west side of the driveway, belonging to Wes and Rose Simchick, whose land we bought a few years later. On June 1, the potato plants would have been low-lying, darkly leafed vegetables, separated by finely cultivated beige soil.

The railroad tracks came next, without any guard or crossing gate. I'm sure I bounded over them, not realizing that the tracks were still in active use by passenger and freight trains that ran on erratic schedules. They looked too abandoned to be dangerous.

After the tracks there was a stretch of farmland still planted in a cover crop of rye grass, which waved in the breeze like the sea. It could have been any farm anywhere. Then, coming up on a rise, I could see the stakes that Alex had set out to mark the new vine rows. This was it! I slowed down to look at where he had planted the first vines, which I knew were Pinot Noir and Cabernet Sauvignon to the west and Sauvignon Blanc to the east. They were almost invisible—just little sticks poking about two inches out of the ground, one every seven feet. My heart started to beat faster. I glanced around, but I still didn't see Alex.

At the end of the half-mile drive there was, first, a garage-sized tractor shed that was encased, like the house and a bigger garage, in chipped white asbestos shingles. To the left was the classic Long Island cedar-shingled barn, very angular, with a sliding front door, a lean-to on the side, and a concrete potato storage building running behind it. These buildings were full of rusted and broken farm tools, piles of old burlap bags, and the greasy spill marks of countless leaky crankcases. I didn't enter them for weeks. It was a whole year before one of us broke through a rotten floorboard in the barn, revealing an old and empty bootlegger's cellar. Later we were told that a prior owner of our farm had been arrested for hiding hooch there during Prohibition.

It took me a year, too, to dare enter the tattered strawberry sorting shed, whose east side had housed a farm worker. The destitution of that long-gone man left its traces in the rat-infested mattress of an old bed. A primitive refrigerator, containing the mummified remains of a dead raccoon, waited unopened until the following summer, when I finally had to go in there and transform the shed into a habitable space for Alex's sister Meg, who returned for the season to help us out.

At last, getting out of the car, my legs wobbling from the long drive, I found Alex out in the field with Meg and Charlie. There were hugs and kisses all around, and I needn't have worried about Iota. Alex was too busy planting to be very interested in a kitten, but he did reach over to pet her little head. With a grin on his face he proudly showed me how he had laid out the vineyard rows in a nine-by-seven-foot grid. He pointed out the precious new leaves that were unfurling from the grapevines' buds, which had turned a brilliant green edged in pink, like a Lilly Pulitzer dress.

Meg and Charlie were staying in a rental house down the road until the planting was finished a few days later, but Alex and I stayed in the farmhouse from the day of my arrival. I was glad that when quitting time came Alex and I would be alone, but as I stood watching them finish their day's work, I felt like an outsider. I didn't know Charlie or Meg well. They were half a decade younger than I was, and I had seen them before only on the occasional vacation. The two were so close to each other that they communicated like identical twins, with looks and gestures being sufficient between them. They baited and goaded each other; they met in a huddle with Alex; they all mumbled and giggled about something I couldn't quite get. I was jealous of these gangly postadolescents, kicking pebbles absentmindedly in the field with a familiarity that made the farm seem more theirs than mine.

I was reassured in a short time, when we heard the village firehouse sound the six o'clock whistle, announcing quitting time for the three of them. I put my kitten down while Alex carried me across the threshold of our farmhouse. He had already carried me across the threshold of our first apartment, where we had settled in as newlywed students at Harvard in 1968, but this time I felt it was for keeps. The house was practically empty, but it was ours. The man who had been renting it had moved out the day before, and our own furniture had not yet arrived out of storage, where we had left it two years before. Having made only a cursory inspection of the house before we bought it, I now took a closer look.

Right in front of the cottage was a pump house. It covered a well that the original settlers must have dug. Now there was a pipe running down the well that indicated it was still in use. Looking down the hole, you could see that it was lined with the red stones that had come from New Haven before the Revolution. The first Long Island settlers had traded flax for these building stones, since their own soils yielded plentiful crops but no stones larger than pebbles. A steel windmill tower rose above the pump house, no longer equipped with the blades that used to pump water and generate electricity for the house. I liked its look of self-sufficiency, even though it was ugly and hazardous.

A small covered entrance to the house led from the front door into a hallway that had hooks for coats. On the immediate left was a small bathroom that had a rusty steel shower stall, so that anyone entering caked with mud or contaminated with pesticides could get clean before going any farther. To cover the chipped wallboard, Alex later hung an art nouveau print that had been in his childhood bedroom. It was a storybook view of aristocrats in sculpted wigs and flouncing dresses, having a picnic in the countryside. The closet in this bathroom was the only closet in the house, and it had never held either wigs or gowns.

When I entered the kitchen, I thought I was in the wrong house. I hadn't paid enough attention when the Realtor had shown us the place, and I hadn't noticed how nasty and rotten the old linoleum was. This room and the dining room next to it were the entirety of the original house that had been built around 1680. Until we dismantled the walls and ceilings later that winter, there was no sign of the beautiful, handhewn beams that supported the structure, nor could we see the original, hand-milled pine boards, as wide as an ancient tree, that lay under the gouged-out linoleum. I would have been much less distressed if I had been able to foresee that within a couple of years, the shabby space would ring with the laughter of our own child, as she zoomed across the old floors in her bright green walker. Nor would I have disparaged the circular fluorescent light that hung over a dented steel table if I had known that before the summer ended, we would sit at that table with our neighbors, sharing a feast of freshly caught crabs whose colors shone in the glow of that lamp.

"Oh, good! A gas stove," I said to Alex, pointing to a sixties-model range in the corner.

"It's good if you like the smell of dead rat," he replied. Apparently, something had died in the insulation. We'd have to get a new stove. I wondered what else we'd have to replace.

The next thing I saw was a crooked chimney, looking precari-ously close to collapse. It rose from a furnace in the basement up through the ceiling. The last family to live there must have taken out the old colonial hearth in order to modernize the heat-ing. It looked as though they had started building the chimney from the roof down and had had to alter the chimney's course when they got inside and realized it wouldn't meet the furnace below. I could see old mortar crumbling from the stress of bad physics.

The doorway that led from the dining room to the living room was the only one that remained from the original colonial structure. Built for seventeenth-century-sized people, it was so low that I could foresee Alex, who is six-six, whacking his head on the lintel. The doorway led to a small living room that the owners of the house had built some time shortly after the Civil War. I did not imagine how bright this dingy space would be-come with our first Christmas tree, decorated with cookies that Iota would bat around until they fell and she ate them. The liv-ing room also had a door that led outside to the back of the house. Later, when we took the ugly pink paneling off the walls, we uncovered the little row of windows known as "lights" that had commonly surrounded the front doors of New England houses more than a century ago. These lights indicated that originally the house had fronted a road that ran along that edge of our property. Now the land only a few yards from our house was heavily wooded. I could see a line of old cedar trees that marked the abandoned traces of that road, and I wished the open road were still there to bring more sunlight into the house.

Looking out the window, I thought about others who had lived here and farmed this land—how it had looked to the early settlers, and why they had chosen the crops they had grown, go-ing back to the time they shared it with the natives who already tilled the soil before the Pilgrims landed.

When the first English settlers came to the North Fork, they would have found tall forests, with small clearings where the natives had planted corn and squashes. Along the shore were sandy inlets with sheltered coves and creeks full of clams, oysters, mussels, crabs, and spawning fish. Historically, most of the farms had grown a little of everything, with some extra dairy, strawberries, and asparagus for the market. An influx of workers from Poland at the end of the nineteenth century had changed the emphasis to cabbages, cauliflower, and potatoes. In the fifties, when the farmers replaced their horses with tractors and bought combines to harvest the potatoes, they had too much capital tied up in equipment for potatoes to afford to grow much of anything else.

Before the American Revolution, the English settlers burned down the trees in the land known as "the Devil's Belt," which ran along Long Island Sound on the North Shore, facing Connecticut. They were worried about the Spanish Armada and wanted to make it easy to see any approaching warships. The fires chased away all the wolves, bears, and rattlesnakes that had been living there. What remained were the eastern deer, squirrels, possums, and garter snakes. I felt relieved, thinking that all those dangerous beasts were gone. The dangers that lay ahead for us—not as cataclysmic as a war but challenging nonetheless—were unimaginable at the time.

Leaving my reverie about the settlers, I went on to inspect the bedrooms. At first, I was annoyed to see that one of the two small bedrooms that came off the living room was completely filled with the worldly possessions of the man who had been renting our house just before we bought it. Alex explained that he hadn't wanted to get on the bad side of this fellow, whose temper had left its mark in various holes he had punched with his bare fists in the walls of the bedroom and the garage. We thought we'd have to store those things for a month at most. In fact, they ended up there for much longer. The only one of us who wasn't bothered by this arrangement was Iota, who loved to hide in the jumble of moldy stuff.

There was a second bathroom in a sort of lean-to off this bedroom. It must have been an exciting innovation when it had been added some time in the fifties, but now its plastic tiles were pitted and dirty. This was the first room that we wallpapered, in a pattern of vines and songbirds. A few years later, when real songbirds threatened our crop, I wished we had chosen paper with flowers or Chinese pagodas.

The other downstairs bedroom had two windows, one to the north that looked out on the barn and one to the east where there was a big old blue spruce tree. Alex chose that room as his study—officially the vineyard office. Charlie built a bookcase along one wall, and along the opposite wall Alex placed a gift from his parents, an old rolltop desk that had once belonged to the mayor of Rochester. I didn't have a room of my own, but I didn't expect to have one, either. Alex was supposed to be the manager, so of course he needed an office. It became a sort of refuge for him, and eventually, after the work of the vineyard piled up and closed in on him, he painted the room saffron yellow. It enveloped him like the robe of a Buddhist monk.

A steep and narrow staircase led upstairs to two more small rooms, each with a single window at the only point one would fit under the sloping roof. There was also an attic, which had been the sleeping loft over the original two-room house. When winter came we removed the attic floor in order to raise the ceiling of the kitchen and found, hidden under the floorboards, a handgun with "1884" embossed on its handle. It was too small a firearm for hunting animals. We wondered what danger had led the family who lived there at the time to buy a handgun. We speculated about husbands murdering wives, or neighbors shooting neighbors. We even entertained the possibility of ghosts, but if there were any ghosts in our house, they didn't reveal themselves to us. Once, when we were staying in the spare bedroom of my sister's pre-Revolutionary house in Connecticut, the ghost of a solemn young man dressed in late-eighteenth-century clothing had appeared in the night. We were ready to meet any lonely, searching specters who might have inhabited this old place, but we never did. Maybe we were too preoccupied in the ensuing years to notice them.

After a disoriented night spent on a mattress upstairs, I awoke early, eager to try my hand at planting. I thought it would make me feel that I was a part of it all. As it was, not much remained for me to do. There were only a few rows left to plant, but at least I got to see what it was like. When we unbundled the plants to trim their roots and sort them for quality, some of the tender graft unions fell apart. I was reminded that it was the very fragility of these plants that had brought us to Long Island.

Every old farm on the east end of Long Island has a grapevine on a trellis between the back door and the outhouse, but that wasn't the kind of grapes we planted. We were defying three hundred years of history by planting *Vitis vinifera*. These are the noble wine-bearing grape varieties that were first planted in Persia and resurrected by Noah after the flood. Just as the Greeks stole Helen from Troy, they also brought the vinifera vines from the Middle East to flourish in their land and beyond. During the reigns of the Caesars, the imperial Romans took these vines from Italy to France, Spain, and Germany. The descendants of these grapes that they planted in Bordeaux (Cabernet Sauvignon, Merlot, and Sauvignon Blanc) and those that they planted in Burgundy (Pinot Noir and Chardonnay) were the vinifera varieties we chose to plant. These varieties have a refinement and elegance that can never be found in wines made from *Vitis riparia* and *Vitis labrusca* (the native American grapes, like Concord or Catawba).

In fact, we found wild native grapes throughout the woods around our house. They are the kind of vines that led the Norseman Leif Eriksson to call the East Coast "Vineland." These native American grapes are coarse cousins of the European vinifera. They grow like tall, ungainly weeds, trying to reach the tops of trees in the forest to get a little sunlight on their big, hairy leaves. American grapes have a strong, grapey aroma, like the smell of Juicy Fruit gum. From the days when foxes lurked around henhouses, these grape aromas were called "foxy." It's not a compliment.

Before we came to Long Island, Alex and I had learned that the early settlers to America's East Coast had been intrigued, then disappointed, by the native grapes. They didn't like the taste of them, and they found that the fruit, with its oddly pulpy "slip-skins," had too little sugar and too much acidity to make stable, palatable wines. Starting in the seventeenth century, from Virginia to New York, vinifera vines were planted and tended, and died. The London Company, a land developer in colonial Virginia, required every household to plant ten vines and learn how to train, prune, cultivate, and tend them. The company brought over indentured French workers to do the work but treated them so badly that the Frenchmen sabotaged the young plantings. In 1662 Lord Baltimore of Maryland planted three hundred acres of wine grapes in that colony, only to see them wither and die. New York's first English governor, the despotic Colonel Richard Nicholls, actually discouraged widespread grape-growing by giving his friend Paul Richards a monopoly on winemaking, including a tax on vines of six shillings per acre a year.

These young vinifera vineyards were all too delicate for the funguses, pests, and cold winters of eastern America. Horticulturists of the early 1800s had some success in toning down the less desirable qualities of the native grapes by hybridizing them, but, while they planted hundreds of acres of these native hybrids just to have something to drink, they still wished they could grow vinifera instead. In 1846, Nicholas Longworth of Cincinnati, known as "the father of American grape culture," complained, "I have tried the foreign grapes [vinifera] extensively for wine at great expense for many years, and have abandoned them as unfit for our climate."

Since growing grapes was going to be a labor of love for Alex and me, we were determined from the outset that we would grow only the kinds of grapes that were used in the wines that had made us starry-eyed newlyweds, especially Cabernet Sauvignon and (most romantic and elusive of all) the Burgundian Pinot Noir. When we learned of the long history of failure with these grapes, we were even more determined to grow them. Without the challenge, we might not have done it at all.

While we did plenty of research to convince ourselves that we could succeed where others had failed, now, as I handled the young and tender plants, I had an inkling that success would not be easy. My first job was to weed out the weakest vines. On every plant is an inch-long piece of wood with a single bud called a scion. The process of attaching a scion from one plant to the rootstock of another is called grafting. The puslike tissue that grows around the graft in order to heal it is what holds the two parts of the plant together. If it's not strong, the plant isn't likely to survive. There wasn't any point in planting vines that had bad grafts, broken buds, or insufficient roots.

If we had been growing native American grapes, they wouldn't need to be grafted. It was actually a disease brought from America to France that made this grafting a requirement for vinifera vines. A reckless nineteenth-century French tourist who visited the extensive Ohio vineyards of the 1850s got very excited about some sparkling pink Catawba wine he tasted there. The Catawba was a grape that was first promoted in the 1820s by Nicholas Longworth himself, and popularized by the American poet Henry Wadsworth Longfellow, who declared:

For the richest and best
Is the wine of the West
That grows by the Beautiful River;
Whose sweet perfume
Fills all the room
With a benison on the giver.

The Frenchman decided he could make a killing producing a similar wine in France. But the native American vines he took back home carried a bug on their roots—a little louse called phylloxera—that did no damage to the woody roots of the American vine but quickly spread to feast on the fleshy roots of neighboring French vinifera vineyards, choking them to death.

After the phylloxera appeared in France, it took years for French scientists to figure out why all their vineyards were dying. First, they pulled out the dead vines and fumigated their soils with poison. Once they realized that phylloxera didn't kill native American vines, they then tried grafting their vinifera vines onto American rootstocks. In the grafted grapevine, the scion, made from the type of vine whose fruit is desired, grows above ground, while the native American rootstock puts out roots in the soil. It took many more decades to discover which varieties of American rootstocks were compatible with which varieties of vinifera. Research continues to this day, in fact, for specific soils and climates. The vines that we bought were grafted, but their rootstock had never been tested in Long Island soils. There was no way of knowing if they would be compatible.

When I looked at the callused plants in Alex's hand, I thought that like the vines that were grafted onto each other, he and I were grafted, too. I wasn't worried that we might not be compatible. By the time we made a serious search for vineyard land, we were used to doing everything together. As Alex showed me the grafts, touching my hand as he passed a bundle of vines to me, I thought, This is what I always wanted.

It felt right for me to be so close to my husband. I always craved that closeness. I recall a tussle I had with my mother back when I was a kindergartner, on the day that class pictures were to be taken. She chose a pink dress for me to wear. As usual, my dress had buttons up the back, with a sash at the waist— a fashion that made it impossible for little girls to dress themselves. The sash was a two-inch-wide strip of fabric stitched into the side seam of the dress—a vestige of the Victorian obsession with bows. My mother did up my buttons, and then tied my sash. "It's not tight enough!" I cried. She tied it again, tighter. "It's still too loose!" I cried. She did it again. Now, it was time to go, and my baby brother was crying, too. As I walked to the car, I could feel the sash loosening. "Fix my sash," I wailed. My mother yanked the sash so hard that it tore right off the dress, and I had to go to school that way. In the class photo, I can see myself, twisted sideways in a loose, wrinkled-looking frock. Years after I was married, my mother told me that she wishes she had just picked me up and hugged me, because that was all I really wanted.

After we finished sorting our vines, we loaded as many of the plants that had passed our inspection as could fit into the bin of our planter. The rest waited in a bucket of water. The planter was like a little surrey that attached to the back of the tractor. It had a tray to hold the plants and two metal seats over a blade that parted the soil. While Charlie kept soaking and preparing the vines and Alex drove down each row, Meg and I worked together, one of us handing a plant to the other, who would insert it into the furrow. Two wheels would then push the soil back over the roots. Immediately, I found out how hard a task this was. Our tractor, a new John Deere, had an exhaust pipe that pointed directly into our faces as we rode behind it. The tractor dealer hadn't bothered to tell Alex, who had custom-ordered this particular machine because the local potato growers' tractors were too wide, that he should have specified an exhaust pipe exiting from a smokestack in front of the engine. Whether the dealer was playing "Welcome, stranger" to what must have seemed to him like a young man pursuing a ridiculous venture, or whether he had made an honest mistake, we'll never know. The net result was that within minutes Meg and I were choking on the fumes. We had to dismount at the end of every row just to breathe some fresh air. This experience was something new that I shared with Meg, Charlie, and Alex, but they had been doing it for weeks. In light of that, I certainly wasn't going to complain.

As we rode on the planter, it lurched and rocked over the lumpy soil. Our planting had started so late in the season that there had not been time to prepare the soil properly. If we had it to do again, we would have plowed and limed the land a month sooner, before the winter cover crop of rye grass got so tall. We hadn't known before Alex arrived in Cutchogue with his load of vines that we wouldn't be able to buy a tractor off a dealer's lot. While he waited for delivery of the custom tractor, he realized that if he couldn't find someone to plow our field soon, we would never have time to get the plants in the ground before they sprouted and died. When he asked the dealer to suggest someone who could do the job, all the farmers who were recommended were busy planting their own fields in potatoes.

Alex called me in Rochester despairing of finding anyone to plow for us. There he was, on his own, with his young brother and sister, trying to figure out how to start this enterprise, and he had never planted anything in his life. How was he going to pull it off when he didn't even have a tractor? Finally, he discovered that we had something that most potato farmers need—a large potato storage barn that was empty. We wouldn't be using it for a couple of years, until we started a winery. Frank McBride, the man who had rented our farmland from the previous owners, had been worrying out loud about how he was going to find another storage building now that these kids from Rochester had moved in and bought this one out from under him. Ever the diplomat, Alex went to him and said, "Frank, if you'll plow our land, you can use our barn until we need it." A hearty handclasp sealed the deal, and the next morning Frank was over with his plow. Alex called me again a few days later, a little tipsy from sharing a bottle of Taylor's sweet wine with Frank after the plowing was done. Frank had brought him the bottle as a gesture of welcome; he had cried when he told Alex how much he loved farming our land, and he hoped we would take good care of it.

Because the soil was plowed so late, the grass had no time to decompose before we planted, so the blade of the planter kept getting caught on rotting stalks. The stalks would throw off the rotations of the coulter, a wheel alongside the planter that measured the distance between plants. Just as I was about to stick another vine in the ground, the grass would accumulate on the blade and we would have to stop, clean it off, and start again, having lost our measured distance between plants.

I planted only for a day. All of the vines we had bought that year, seventeen acres of them, were now in their designated rows, putting out new rootlets in the soft soil. The tension I had felt in my jaw had been replaced with a new feeling, one of total physical fatigue and mental exhilaration. Alex and I embraced as we celebrated the completion of planting and the beginning of our life as vintners.

The following day, after a night of sweltering under the eaves of our upstairs bedroom, we elected to move our mattress outside while we tore down the upstairs walls to make a single bedroom there. In the same way that planting the vines had tied my physical self to the earth of my new home, sleeping outside wove the location into my psyche. Each night, my bare feet trod on dewy grass as I made my way to the large mattress we'd placed under our maple tree. The leaves above us made a pattern of cut-work design that was especially beautiful in the early dawn, when the air was completely still and the sky behind the leaves looked like handblown cobalt glass. The grasshoppers played a stacatto tune while owls announced the night, and sometimes we could hear the shriek of a fox. Soon, the lilac hedge bloomed, mingling its fragrance with that of the grass and the newly plowed field. We knew the phases of the moon and saw the Milky Way on nights when it wasn't obscured by humidity. The night sky was also filled with the constant blinking of aircraft going from New York to London or Paris, or left in a holding pattern, as seventy miles to the west the air traffic controllers at Kennedy Airport dueled with the controllers at La Guardia Airport for airspace. In August we moved the mattress into the open so that we could watch meteor showers pierce the slow-moving tracks of these planes. Rain showers sometimes sent us back indoors, dragging the mattress, but they happened infrequently during the long, dry summer.

Iota loved to pounce on us in the mornings. She was becoming a sleek and wily cat, as loyal as a dog and just as demanding. Her Siamese blood revealed itself in her ability to make her every desire known. She expressed herself with humor, like Alex, and it didn't take long for him to appreciate her intelligence. The few white whiskers she had over her eyes made her face expressive. If she was worried, she'd twitch them. If she was relaxed, they would droop as she slowly lowered her eyelids until her green eyes were like slivers of a newly waxing moon. I could tell her mood by how she switched her tail—slowly like a hula dancer when she was contented, or stiffly like a drum major's baton when she was on the prowl. She knew how to get my attention by making figure eights around my bare legs, touching her fur against my calves like silk pajamas.

Naturally, as time went on she learned to hunt and left

us little presents by the front door. She knew enough not to bring them inside. The presents were usually dead moles, because if she caught mice, she'd eat every trace of them herself. I did a little research and read that cats don't eat moles, because they taste bitter. The enterprising scientist who came up with this theory must have interviewed a lot of cats, or tasted moles himself.

After the planting was done, we went into town to buy groceries. In 1973 the village of Cutchogue consisted of a single block of stores. There was a post office, a dingy pharmacy, a "variety" store with inventory that hadn't been touched since the 1930s, and a small grocery store that smelled of rancid kielbasa. What the town lacked for shopping, it made up for in churches. To serve a population of under three thousand souls, there was a Methodist church, a Presbyterian church, a community church for the folks who split with the Presbyterians a century ago (now used as a library), and two Roman Catholic churches—one for the Polish and one for the Irish. We kept on driving two miles west to Mattituck and saw another church for the English Catholics. Mattituck also had a new, modern supermarket. In those days, stores were closed on Sundays, and before I arrived Alex had discovered that if he shopped near closing time on Saturday, he could bargain with the bakery department in the new supermarket for pastries that would otherwise be thrown out. He liked the mocha layer cake—doubly good at half the price.

I was eager to get started with my own baking, so we soon threw out the stove with the rat in the insulation, and bought a six-burner Garland range with two large ovens and a griddle. That restaurant stove became the heart of the house and the means to comfort, nurture, reward, and inspire anyone who entered the vineyard and became part of our great experiment. What is wine without wonderful food? Who would want to toil in the fields without the reward of an ample supper? I wanted to be the big farm mama, making big farm meals. In the morning, I would pick wild mulberries off the weeping mulberry bush in our front yard. I'd blend an egg, a cup of sour cream, some vanilla, and a little sugar with a cup of flour and a teaspoon of baking soda, throw in a handful of mulberries, and thin the batter with a shot of milk. When the griddle was hot, I'd bake batches of pancakes that Alex and I devoured. You can eat any amount of pancakes when you're a farmer.

However much I enjoyed cooking at home, that first summer in the vineyard we ate most of our suppers with our neighbors, Mike and Irene Kaloski. Within days of our arrival on the farm, Mike, a thin, balding man in a long-sleeved shirt and faded trousers, had come poking around to see if the rumors he'd heard about a couple of kids with a cockamamy vineyard scheme were true.

Where I had grown up, sixty miles west of our vineyard in the Long Island suburb of Cold Spring Harbor, neighbors pretty much kept to themselves. My family lived next to a cemetery on one side—not a lot of conversation there. There was another family in a wooded area on the other side of our house. They were friendly with us and they even said they'd let us share their bomb shelter if there was a nuclear war, but otherwise we all minded our own business. When Alex and I looked for property on the North Fork, we didn't think about what our neighbors would be like—all we cared about was the suitability of the land for grapes. We were completely consumed by the task of starting our vineyard. We hoped that if we didn't bother them, they wouldn't bother us, like pioneers and Indians.

It came as a complete surprise when one of our new neighbors gave a small reception for us, which they called a "social tea," to introduce us to some of the congregation at the Presbyterian church. Alex and I had both been brought up as Episcopalians of the most lax variety, but I had many ancestors who were staunch

Presbyterians, so that wasn't too great a stretch. Even so, I felt as if we had traveled back in time, to somewhere at the beginning of the century, as I stood in our neighbors' parlor that hadn't been changed for a hundred years and shook hands with those teetotaling old Yankees. They stood at a careful distance from one another, balancing their teacups and raising their eyes to meet mine for only split seconds at a time.

Time would prove that our neighbors in Cutchogue would help us in ways we could not have imagined. Of them all, Mike was the most helpful in the first and most critical year that we farmed. He was a sort of pioneer himself, because he had dared to go into territory where he didn't belong, and because he planted crops that everyone said he couldn't grow. Mike spoke like an engine that has just kicked in, with hesitant pauses that were followed by rapid-fire assertions. He had lived all fifty-odd years of his life on the same farm in Cutchogue, with Polish parents who never learned to speak English. Since most of the other farmers in town were also Polish, they didn't need to. He had learned English in the one-room schoolhouse that still stood as a historic curiosity on the Village Green in Cutchogue.

One day when Mike was in his mid-twenties and still single, he went to the movies and met a beautiful girl named Irene. When he asked her for her address, he realized he had a problem. Irene lived in New Suffolk, a fishing village only about a mile from Cutchogue whose residents were of either English or Irish descent. They didn't want to have anything to do with the Polish. But Mike was smitten with Irene. So he went down to the bar in New Suffolk. He bought a round of drinks for all the men, and then another round, and by the time he let on that he was there to see Irene, they all thought he was a jolly good fellow and let him get on with his courting.

After Mike and Irene got married, he couldn't afford the rounds of drinks anymore, so he wasn't welcome in New Suffolk, and she wasn't welcome in Cutchogue. This went on for many years, until they had children, which opened everyone's hearts and doors. Maybe that's why they knew how to be good neighbors— they had been snubbed themselves.

Mike loved farming, and he loved trying anything new. Every winter, he liked to peruse the seed catalogs, and he would order things that weren't supposed to grow in our zone. He grew peanuts, which Irene canned as if they were chickpeas. When we went fishing one day on the motorboat Mike shared with his brother, Chet, they treated us to these boiled goobers, washed down with blackberry brandy. On the boat, we discovered that neither Mike nor Chet could swim—another legacy of having grown up Polish in a place where the beaches were the province of older, English stock.

Watermelons were another marginal crop that Mike grew. The big, southern-style melons didn't ripen well, so he tried the round icebox melons, which he'd share with anyone who stopped by his house when he and his family took a break from harvesting potatoes. Rivulets of sweet red juice from the melons streaked their dusty faces as they bantered and teased each other, sitting outside their cottage in a tight circle of nylon-webbed chairs.

We could also find them in the same circle of chairs when they sorted their crop of shallots. Mike was the first to plant shallots on Long Island, and it took him several years to find a source of seed for the little onions. Other growers earned such a premium on shallots that they did not want more of them on the market.

Mike had been raised on whatever meat his family could hunt, and he still hunted rabbit and duck. There were plenty of those in the fields and marshes. Every morning, I enjoyed the hauntingly sad cooing of the mourning doves that fed in our fields, so I was taken aback when I stopped over at Mike's one lunchtime and was offered mourning dove soup that he had made himself. It was a clear broth in which the headless doves floated, with their fat legs sticking up and their pale skin, bumpy from having just been plucked, shimmering in the surrounding liquid.

From the first day that Mike wandered over to our farm, he was our mentor. Even so, we didn't always take his advice. My introduction to the danger of weeds began as a warning from Mike—a warning we didn't heed until it was too late. I recall walking into the vineyard that first year, just a few weeks after the grapes were planted. The soil looked nicely cultivated to me, but Mike wanted to show me something. Bending down, he pointed out the primordial leaves of some reddish stuff that looked like an inconsequential dusting of color. "Red weed," he announced. "Now is the time to get rid of it. And this"—he pointed to a fleshier sprig—"is puzli." He meant purslane.

I learned the names of dozens of weeds because they became such a familiar menace. Some of them, like the purslane, dandelion, sorrel, chamomile, lamb's-quarter, and pepper weed, were edible. But the definition of a weed is a plant that grows in the wrong place, so it didn't matter how desirable it might have been under other circumstances. In the vineyard, these interlopers would compete with the vines for moisture and nutrients, and promote the growth of bugs and funguses.

We hadn't thought about weeds as such an imminent threat, and it took a few weeks for us to get a cultivator set up for the size of the vineyard rows. Meanwhile, the weeds became taller than the vines. Even after we cultivated the rows, until the vines were mature the only way to control the weeds under the trellis was by hand, with a hoe. Alex found that hoeing was too hard on his back, but my back was strong, so I went out into the field and started whacking away at them. Never in my life had I thought I could spend seven or eight hours a day repeating the same motion over and over. For a while I fought it, but out of necessity I learned to let the hoe blade and gravity work *with* my muscles. This motion would feel good, then it would hurt, then I would just go elsewhere in my mind as my arms went numb and my brain went on automatic pilot.

I spent most of that first vineyard summer hand-hoeing seventeen acres of vines. I did it three times, by myself, and I ended up doing it for part of every summer after that—like the mythical Sisyphus, whose eternal punishment was to roll a boulder up a hill and watch it roll back repeatedly. I didn't think of it that way then.

2

Farm Labor

ONE DAY IN JULY, MIKE CAME LOOKING FOR US IN THE MIDDLE of the morning and found us still in bed. Trying to control his astonishment, he slowly turned his head back and forth as he considered what to say. Then, putting his hands on his hips to show he was serious, he said, "You're farmers now, and you've got to live by the farmer's clock. What do you think those whistles at the firehouse are for? Get started at six. Eat lunch at noon. Work until six, and then eat supper. Rest on Sundays unless there's an emergency."

From then on, that's exactly what we did. Back in the field, our baby vines were putting forth lush growth, with long green canes that trailed over the ground. When July brought a drought and our plants started to shrivel up we began to panic. Our 1940s-era Buda diesel irrigation motor was broken, and we didn't have irrigation pipe anyway. Mike, who was irrigating his potatoes twenty-four hours a day, moving the thirty-foot-long sections of aluminum pipe every four hours, came by to see how he could help us. He told us that we could probably buy some pipe at a farm auction that was taking place that week. So many potato farms were going out of business at that time that these auctions took place frequently. Meanwhile, he worked on the Buda. It took some tinkering, but he got it going. Looking at the jets of water circling over the wilted vines, he said with satisfaction and a twinkle of irony, "There you are. Mike Kaloski can irrigate better than God."

At the farm auction, we met the second man to plant wine grapes on Long Island, David Mudd. We were standing in a parched farmyard, surrounded by rusting equipment that only a few remaining farmers had any interest in buying, when we spotted a heap of irrigation pipe. As we hung around waiting for the auctioneer to get to the pipe, we struck up a conversation with Dave, a square-faced man with short, curly hair, a genial smile, and a Kansas accent. Dave was a pilot for Eastern Airlines. Like other pilots, the time he could spend flying was restricted. He was looking for an interesting project that he could do in his spare time. Since he had grown up on a farm in Kansas, he thought he might do a little farming on some land he owned in Southold, about five miles east of us. He told us that he was planning to grow fancy grasses to feed the horses that were going to be at a proposed racetrack nearby. He hadn't planted any grass yet; first he wanted to buy some used farm equipment, including that irrigation pipe that we planned to bid on. We told him about our vineyard and described how perfect we thought the North Fork would be for specialty wine grapes. By the time we had outbid him for the pipe, he had changed his mind and decided to grow grapes, too. We agreed to supply him with his first vines, giving him part of our order the next year so that he wouldn't have to wait to order his own plants. Dave became a regular visitor to our farm, coming by to learn as he lent a hand weighing pruning wood, testing different tying materials, and generally scoping out the scene. Within a few years he had started a business planting and tending vineyards for investors.

We looked forward to Dave's visits because he told stories about his piloting adventures. He told us of how he had been trapped on the ground in a burning aircraft, and his description of how he had gotten his body through a window that was narrower than his hips was proof enough for me that he was someone who would not give up on anything easily.

Right around the time that Dave started his vineyard, there was a rash of airplane hijackings. Dave flew the route from New York to Puerto Rico, and many of his passengers had never flown before. They would get drunk as quickly as they could as soon as the plane became airborne. "One time," Dave related, "the stewardess called me from the cabin, saying that a drunken passenger had taken out a knife. I immediately dropped the aircraft, and by the time I leveled off again I had another call from the stewardess saying, 'He's put away his knife and taken out his rosary beads.' "

When the autumn of 1973 came, Mike Kaloski invited us to make wine with his father and sisters. They had been making it since Prohibition, when selling jugs of homemade wine and whiskey for eight cents a gallon had helped to feed their large family. That put them a step ahead of us, since we had never made wine at all. We didn't have any grapes yet, so we used the elderberries that grew wild in the hedgerow that separated our farms.

Elderberries grow in a starburst of tiny black berries that have to be stripped off their stems individually. Although elderberry wine can have depth and structure akin to a fine Cabernet Sauvignon, the berries are too bitter to eat raw and too much trouble to separate from their stems for most people to bother. The Kaloski family had grown up at a time when anything edible was precious, even if it was hard to make into something worthwhile. They worked as a team, and it never seemed to be less than a lark to work with them. Mike's sister Cookie had long ago moved to Queens, but she was there for the season, mostly to help Mike pick beans. She relished the chance to make a little wine. Another sister, Yippie, drove a school bus during the winter but was ready to lend a hand as well. Their children came along as a matter of course.

First, we gathered quantities of the berries, carrying them in well-scrubbed, empty plastic buckets that had originally held pesticides. We sat on overturned bushels and raked the berries off the stems with our fingers, filling an old barrel as we chattered and teased one another. As we covered ourselves in inky juice, we took turns stomping on the fruit before adding water and sugar to the must. We had to increase the sugar and decrease the acidity or the wine wouldn't be drinkable. Mike's eighty-year-old father led us in singing a Polish song, which sounded like *"Yesta polska nie zignela, poki me ziami!"* I figured it had something to do with drinking and polkas. Actually, it was the Polish anthem, and it meant "Poland's not dead as long as we live." Alex and I also led the group in a phonetic version of a French drinking song: *"Chevaliers de la table ronde, goutons voir si le vin est bon,"* which means "Knights of the Round Table, let's taste to see if the wine is good." It was funny to think of this barefoot group as a bunch of Napoleonic-era soldiers or medieval knights. But however different we were, we carried on the same tradition of friendship and loyalty.

We added bread yeast to get the fermentation going and checked on it over a few days until there wasn't any more foam on the surface of the barrel. Then we pressed the spent berries through cheesecloth and put the new, dark wine into glass jugs. If that was all there was to winemaking, I thought, it was easy. We wouldn't have a problem when it came to fermenting our grapes.

Mike made everything seem easy. We were lucky to have his help. Although we were full of bravado, the fact is that we had to learn everything from scratch. Ours was a total do-it-yourself operation. We took it step by step. In one field rabbits came out of the nearby woods and savagely ate the Sauvignon Blanc vines back to the ground. We found a dairy that could supply us with milk cartons, and put one around each vine on four acres to save them from the little monsters.

The fact that the vines grew at all was a miracle to me. It was a relief to see the enormous size of the vines by the end of the first growing season. Pleased to see that the rootstocks were obviously compatible with the soil, we weren't aware that they were *too* compatible. Their excessive vigor became a problem later on, as the vines produced too many shady leaves, making it hard for the fruit to ripen until we learned to tame their growth.

While I hoed the weeds, Alex worked hard at planning our finances. The money we had invested in the farm came mostly from a legacy he had inherited from his grandfather. Because of that, and also because as a young wife I had never expected to be in charge of making decisions about money, I was happy to leave this kind of work to him. Just as he had bargained for layer cakes at the market, he was good at bargaining with purveyors of farm equipment, building supplies, and, when the time came, the pumps, presses, and tanks needed for the winery. When it rained, we would work together on fixing up our house. Sometimes, while I read scientific treatises on winemaking, Alex studied the latest equipment catalogs and agricultural bulletins from Cornell, or read the sort of historical and poetic texts that fired his interest in farming, like Virgil's *Georgics,* or Ulysses Hedrick's 1908 book *The Grapes of New York*. We got used to working, so that when a day dawned warm and sunny, I no longer thought as I had early in the summer, Let's go to the beach.

As our first year in the vineyard progressed, we got to know more of our neighbors. Alex had the idea that we could invite the Kaloskis along with some of the other people who lived along Alvah's Lane (the road that ran alongside our vineyard) for a cheesecake contest. We supposed that people in this farming community were likely to be good cooks. Besides that, Alex loved cheesecake. Everyone was to bring a homemade cheesecake, and we would provide beverages—lemonade for the teetotalers, and French Sauternes, the sweet white wine of Bordeaux, for the rest of us. No one had ever been invited to a party like this before, but it was a great way for us all to get together. About twelve people came, yielding six or seven cheesecakes. I think most of our neighbors came just to see what the inside of our house looked like. I made my favorite cheesecake, a simple recipe made with cream cheese, eggs, sugar, and vanilla. The other cheesecakes ranged from the fluffy kind made from Jell-O to a spectacular one that was dense like mine and covered with homemade brandied peaches. Everyone tasted and voted on the cheesecakes; the one with the peaches took the grand award—a bottle of Sauternes. Our neighbors who were Polish weren't used to socializing with those whose ancestors had come to town in the 1620s, but we all loosened up and enjoyed the party. The next time we saw each other at the post office or on the road, we had something to talk about. The contest was never repeated, but it endured in local legend for years.

Now that the planting was done and we had a big stove and a guardian cat, it was time for us to start a family. We had already been married for five years. At last we had our own home. Anyone else would have seen that this was a stupid time to introduce the responsibility of caring for an infant, but to me, having a baby was part of the pioneering dream. Even though I still planned to continue my studies in chemistry at the State University at Stony Brook, forty miles away, I thought I could do everything I wanted to do—school, vineyard, and children—at once.

I thought of the vineyard as a project—my life's work. When

it was just a few acres of unruly plants in competition with a bunch of weeds, I didn't think of the vineyard as "my baby"—something I would nurture like a child. That was Mother Earth's job. For me, giving birth would help to define who I was. I wanted to be a farmer, and there was no question in my mind that being a farmer meant having babies.

After the August meteor showers we moved back upstairs under the roof, and by November I knew I was pregnant. I loved being pregnant. Instead of having morning sickness, I only felt stronger and healthier. When winter came I helped Alex take a sledgehammer to the clumsy walls that made our house feel cramped. Together, we put up new Sheetrock and spackled and painted the walls. Outdoors, we began to prune our vines. We brought our textbook into the vineyard. The flat black-and-white photos of vines didn't look much like our tangled masses of growth, and it wasn't as easy as we thought it would be to decide which canes to keep and which to cut away from the base of the plants.

Later that same winter, in February 1974, a posse of professors arrived to inspect our young plants. In the lead was Dr. Nelson Shaulis, a professor of viticulture at Cornell's state-funded research station at Geneva, New York. Shaulis had dedicated his life to improving eastern viticulture and had some highly important theories about trellis management. Along with him was Dr. Amand Kasimatis, another professor, all the way from the University of California. It was an honor to have these two important men come all this way to help us. Surely they would be able to show us how to prune our vines correctly.

We took the two men out to look at our little plants, which at this point were completely dormant. Having lost their leaves for the winter, the vines showed off their well-hardened wood and substantial vigor. To see if there were the signs of winter injury to our plants that he found upstate in similar vinifera varieties, Shaulis started cutting away at their buds with a razor that he carried in his pocket. In a strangely frenzied way, he slashed, slashed, and slashed until he inadvertently slashed his fingers. "These buds are all green," he declared. His finger still dripping blood, he finally found a wizened bud at the end of a long shoot. "There, you see! Winter kill!" he announced.

Kasimatis was not impressed. "That's just the end of a sucker," he said. "These plants all look good."

I could feel some tension between the two men, but I didn't understand it. They sparred again about how to prune the plants. Both agreed that it was important to cut back the excess growth of the plants now, during the coldest part of winter, when there was no sap flowing that might carry disease to the pruning wounds. One, however, insisted that we should keep only wood that was pencil-sized, with no lateral shoots. The other was certain that the best wood was larger, and that we should keep a few buds on the laterals.

Clearly, we had to make up our own minds. We didn't find out until much later why Dr. Shaulis had been so fervent in his disagreements with Kasimatis.

Another important visitor to our vineyard was Dr. Konstantin Frank, an elderly Russian grape grower from the Finger Lakes who made it clear that whatever Dr. Shaulis had said, he did not agree. Dr. Frank was one of very few upstate growers who planted vinifera. He was excited that we were growing European wine grapes, but, since we had visited him before coming to Long Island, he wanted to take the credit for our new planting. He wandered from plant to plant, stooping over to make his inspection as if we were trying to fool him. "Yes," he announced in broken English, "is Cabernet Sauvignon. Yes, is Pinot Noir." He drew himself up, looked around to make sure we were listening, and said, "You could never have done it without me."

It was clear that Dr. Frank felt threatened by the youth and stature of Alex and avoided talking to him. When he saw me nodding happily at his pronouncements, with my pregnant belly bulging out of my beat-up overalls, he said, "I like you. You nice dirty farmer."

That was just what I wanted to be. After I finished one semester of my course in organic chemistry, I abandoned my formal studies to devote myself wholeheartedly to my farm and my family.

In the spring of 1974, when I was seven months pregnant, we bought another ten acres' worth of grapevines. This time, we bought them from California. A Cornell Cooperative Extension agent who had taken an interest in our vineyard (although he was an expert in vegetables, not grapes) had learned that New York State was planning to implement a virus certification program for grapes that would copy one currently in force in California. He told us that unless we bought plants that were certified virus-free from California, we might never be able to propagate or sell our own plants.

We weren't sure it was all that important for us to buy certified virus-free stock. We knew grape viruses couldn't harm humans, and we knew that some of the world's most prominent vineyards were infected by them. It was true that the viruses could change the color of the plants' leaves or make them curl, and that would interfere with grape ripening by limiting photosynthesis. Leaves that aren't green lack the chlorophyll needed to convert CO_2 into sugar. Nevertheless, some Europeans thought that the fruit of virus-infected vines could be more complex, especially if the viruses were mild.

It wasn't for us to question the merits of the certification program—we had to be in it. The American wine establishment believed that cleanliness was next to godliness. Scientists in California had developed a way to purge plants of virus at the same time that they were being propagated. Tiny cuttings of vine tissue would be exposed to steam while they were being rooted in the greenhouse. At that time, in the early seventies, there was a grape-planting boom going on in California, and everyone wanted these "mist-propagated" vines. The only way we could get them was to buy greenhouse-propagated plants that were only a few weeks old, not the one-year-old, field-rooted variety we had bought before. Ironically, and devastatingly for the acres of California vines that were later hit by phylloxera, the nurseries convinced growers that freedom from virus was more important than phylloxera protection. These heat-treated plants couldn't be grafted in the time needed to get them ready for market. Our business plan depended on our getting our acreage planted within the next few years, so we didn't feel we could wait for grafted, virus-free vines, either. Although having grafted vines had seemed imperative the year before, the nurseries now persuaded us that phylloxera wouldn't live in sandy soils like ours.

Alex and I studied the weather statistics for the average last-frost date, which was May 7. We placed an order for ten acres' worth of the little potted plants to be airfreighted into Kennedy Airport on the following day. The grafted, rooted vines that we had planted the year before had been dormant when we got them, so they were not nearly as tender as these foot-tall, leafy plants. We weren't even sure that these new vines would survive the cold temperatures inside the belly of the airplane.

At least this time we had prepared the field where the vines would go. In April, our friend Paul Kaloski, Mike's son, had taken time out from planting potatoes to drive his big, three-bottomed plow over to our farm. In a couple of days, he had turned the green sea of rye cover crop into a smooth, loamy brown surface.

The first time the field was plowed for our original planting in 1973, I had not been there to see it. This time, as I watched the plow travel down the field, I wasn't prepared for the way the plow blades ripped through the rye grass. It was violent—the necessary deflowering of virgin earth. I watched the dark soil fold over the long grass and saw beetles and worms scurrying and slithering back into its depths. Maybe because I was pregnant, I was especially aware of the life teeming within the earth. I chuckled, and my mind played a riff on worms: a can of worms, worming your way out of something, as the worm turns. I recalled biting into a candy bar when I was seven, at a Doris Day matinee movie, and after eating half of the candy realizing that it was crawling with worms. Turning my back on the golden-haired Doris Day as I raced down the dark aisle to hurl the offending candy in the trash and rinse my mouth out, I did not dwell on the ecology of sharing our planet with worms and insects. Now, I gave thought to how much we needed the help of this underworld of creatures, for the aeration of our soils and the health of our vines.

This plowing seemed to me to be an aggressive act. Who were we to disturb the natural assortment of plants and animals on this large stretch of land? We will work together with the soil, I thought. It spun my head around to consider that modern farming, the kind we were doing, is an arrogant gesture. Smelling the soil, I felt subdued and connected to it and whatever was within it. I walked barefoot in the new furrow and thought that my relationship to my husband was like my relationship to the land—complex, sort of slippery, primitive and deep.

Early in the morning on the day that the new vines were flying to us from California, Alex stood by the telephone in his study, his command post, calling the vine nursery and then the airline to track our shipment. The arrival time kept changing, and I paced the floor as if it were my own child I was awaiting. It was not until late afternoon that the vines were finally on their way. Alex went alone into the quagmire of Kennedy Airport's freight system, concerned about how to get the eight thousand vines safely home.

It was late at night by the time we unloaded the cargo onto the lawn next to the house. We could have put the vines into the potato barn, but the leaves would need sunlight during the days that it would take us to get them planted. It would be easy to keep them watered near the house, and the building would shelter them from the wind.

The sky was completely clear that night, but we hadn't been farming long enough to think about radiant heat loss—the way the earth gives up its day's accumulated heat if there are no clouds to cover the sky. Awakened by a full moon after just a couple of hours' sleep, we decided to check the temperature outside. It was thirty-three degrees and falling. We grabbed every blanket and sheet and piece of plastic we could find and covered as many of the vines as we could. By five o'clock, there was a film of frost over everything. By eight o'clock, some of the exposed leaves had started to turn black. We took the covers off the plants, then put them back on when we realized that the clear morning sky would act like a mirror, intensifying the sun and hurting the cold-shocked plants.

By the end of the day, we were able to rejoice that most of the baby vines had survived. The hundred or so that hadn't, we put philosophically into the trash, and I made a note on the calendar that the last-frost date wasn't May 7 after all.

To get these vines into the ground, we couldn't use the planter that wheeled behind the tractor. Each one had to be planted by hand. Because his spine had been fused, Alex's back wasn't up to the kind of bending required to do the task, and

Meg and Charlie had returned to school. I recruited two young women, a niece of Mike's and her friend from a neighboring farm, to help me.

With the second planting, we put long sheets of black plastic on the ground to keep the weeds down while the vines established themselves. We didn't want to be in the same situation we had been in the year before, losing the weed war. Besides, there were still the original seventeen acres of vines to tend. We were pretty sure we could get ahead of the weeds on those acres this time, but what if we didn't? Black plastic mulch was a new innovation that was being used extensively for vegetable crops in Florida. We hadn't heard of vineyards using it, but we couldn't see why it wouldn't work. It was hard to put down without the proper gizmo to roll it out and tuck the sides under the earth so it wouldn't blow away, but Alex rigged our own roller and we tucked the plastic in by hand.

The main thing I remember about the rest of the planting was that I was too big in the middle to bend over and breathe at the same time. I would double over to dig a hole with a hand-held bulb planter, a metal cylinder with a wooden handle. Coming up for air, I'd hold the square pot with each vine upside down, catching it with my left hand as it fell out of the pot. Holding my breath again, I'd bend down and put the vine in the hole. I think it took another breath to tamp it down. About every half hour, I would have to head to the bushes in order to pee.

This was truly the Zen of grape planting. Definitely yogic breathing. Once I got a rhythm going, it was strangely exhilarating. Breathe in. Dig. Breathe out. Plant. The hard part was talking, which we did all day. I have no idea what we said. The intimacy of working together this way made the topic of conversation irrelevant. We all thought it was fun, and funny.

It was less fun, and not at all funny, a few weeks later when the black plastic was pierced by nutsedge, the only weed that grows everywhere on earth, including the Arctic Circle. We had never heard of it, but Mike explained that it grows from a "nut" that lies a foot below the surface, making it virtually impossible to eradicate by ordinary tilling. Mechanical cultivation would just spread it. Nutsedge has foliage that is stiff and sharp. It can grow through concrete, so plastic mulch was no barrier. We would have to use weed spray to kill it and somehow protect the baby vines from the spray at the same time.

We came up with the idea of bagging each plant with the kind of bag that comes in big rolls in the produce section of the supermarket. We had no way of knowing if the bags would be permeated by the spray, but we couldn't think of any other solution. It was good that Alex had done all that bargaining for cake at the local supermarket, because now the people who worked there knew us. I begged the produce manager for enough rolls of plastic bags to cover three thousand vines. Somehow he had enough bags in stock and sold them to me at cost.

With Mike's help I recruited some of his relatives and a few neighbors to help cover each little plant. Paul did the spraying with a boom sprayer that extended over several rows at a time. Once again, we failed to think through all the consequences of what we were doing. What we hadn't reckoned on was that as soon as the spraying was done, the vines started cooking inside the bags. We had to rally everyone back to remove them.

It took two more days to see that the spray actually did kill the weeds—for the time being, anyway. As the leaves of the nutsedge turned from yellow to brown, we kept inspecting the green leaves of the vines. At first, they started to turn yellow, too. We thought that we had killed them all. Then, they came out of their swoon, brightened up, and started to push out new leaves.

By the time these vines were planted, I could barely waddle up and down the rows. My knees were shot—I never again kneeled when I worked—but the hoeing I had done the prior summer had made my back and arms strong. The overalls I had worn every day throughout my pregnancy were now so stretched out that I couldn't do up the side buttons, so I sewed myself a sleeveless yellow dress printed with little flowers, and a plaid outfit of shorts beneath what looked like a very short sundress. I called the outfit my "playsuit," but I wasn't playing. I was still out working in the field.

After planting, I thinned the extra shoots off the new vines and took shoots that would be below the trellis wire off the older vines, while Alex and his brother Charlie, who had returned for the summer, installed a trellis system for them. They dropped locust posts off of a trailer, setting one post for every three vines, and used a mechanized post pounder mounted alongside the tractor to whack the posts into the ground. For hours each day, I heard the rhythmic *Bam! Bam! Bam! Bam!* as the weight of the machine came crashing down onto the top of each post. It was a nerve-shattering job, and I was glad someone else was doing it.

Eventually, Charlie and Alex strung three sets of galvanized wire down the lengths of the rows, to support the growing vines. They also tried, with marginal success, to stay ahead of the weeds by creating a hill of bare soil over the roots of the vines, using a retractable hydraulic blade mounted on the tractor, and then taking the hill away when weeds covered it. I could tell that by August, after our baby was born, I'd be back out there, hoeing around the posts where the tractor couldn't reach the pesky weeds.

Once a month I visited a group of obstetricians twenty-eight miles away at Southampton Hospital, where I planned to give birth. When I told my obstetrician how hard I was working, he did not discourage me from continuing in the field. With the same serious expression on his face that male doctors make when they are inspecting the delicate parts of women's bodies, he looked at my hands and touched them as if my calluses were especially fascinating.

My due date was fast approaching and I did not dare to make preparations for it. My mother and some of her friends had a baby shower for me, and my sister gave me the bassinet that she had used for her two babies, but the blankets and little outfits that I'd been given sat half-hidden in the crawl space under the eaves of our bedroom. Sometimes I would look at them and wonder if I would ever use them. Was it possible that in a few weeks' time we would successfully negotiate the perils of childbirth?

I had no close friends of my age in the community, and although I gabbed with the other young pregnant women in the waiting room of my obstetrician's office, I didn't feel I knew them well enough to share my anxieties. While my parents lived only an hour and a half away, my mother was preoccupied by the demands of caring for my father, who had multiple sclerosis. I didn't want to put another burden on her. I felt the same way about my mother-in-law, Betty Hargrave, too. She already had enough to contend with since she and Alex's father were on the verge of a divorce. And anyway, I didn't think that our mothers' experience with childbirth would apply to ours. Both my mother and Alex's mother had given birth at a time when obstetricians preferred to anesthetize women and deliver babies by forceps. I wasn't going to let anyone do that to my baby.

There was something else, too, that held me back from asking my mother and mother-in-law for help. Just as the vineyard was our project and ours alone, this baby was going to be uniquely ours.

I have a feeling that this kind of labor won't be like farm labor, I thought. I wanted to attend Lamaze classes in natural childbirth, but Alex didn't. When I think about it now, he was probably just as scared of childbirth as I was, and didn't want to deal with it until it was really happening. I recall his saying that he didn't want anyone to tell him what childbirth would be like; he wanted to experience it for himself. It was understandable for him to want to go into this experience without any preconceived notions. I, however, was the one who would have to push this baby out. I needed more information. Thinking, maybe mistakenly, that he wouldn't want to see the pictures or read about it himself, I got a book about natural childbirth out of the library when I went out to buy groceries and speed-read it in the parking lot. When I got home I decided that I really needed to include Alex and have him help me practice for the upcoming event. I described what I had learned about breathing techniques to him, and he helped me practice them right away.

I got another lesson in childbirth in an unexpected way. I was three days overdue when Eeper, a second cat we had adopted that spring, went into labor herself. Eeper's full name was Marilyn Eep because she acted very sexy like Marilyn Monroe, and because she had made a sound like *eep* when we found her, abandoned as a tiny kitten. She was too young to have babies, I thought, but she must have gotten knocked up on her first date. I was worried about how she would handle childbirth. She was bulging at least as much as I was, and I kept asking her, "Are you ready?" We had a little contest to see who would go first, Eeper or me. When she disappeared, I went looking for her and found her in the crawlspace under the eaves of the upstairs bedroom. I watched her measured breathing, her panting, and her pushing. When the first little bundle emerged, squirming, she dutifully cleaned it and got ready to push out the next. The lesson was repeated six times, and six kittens emerged, each in its own sac. I was really glad that Eeper had gone first.

The next day, as I was tying vines out in the field in my playsuit, I felt a sudden gush of wetness between my legs and realized that my water had broken. Knowing that it was just a matter of time before I would go into hard labor, I went back into the house, tidied myself up, and got Alex to keep me company while I waited for contractions to start. The U.S. Senate was holding televised hearings to impeach President Richard Nixon, so we settled into the living room to watch them, thinking that the vituperation of politicians would distract us. It did, up to a point. By midnight, I still wasn't in regular labor, but the hearings were far less interesting than the prospect of giving birth, so we drove to the hospital.

Southampton Hospital is on the other side of Peconic Bay from our home in Cutchogue. As we crossed the bay, I felt as if I were in a surreal *Twilight Zone* episode. Piercing the darkness, our headlights showed us just a glimpse of our surroundings, as though the only reality were the tiny world of our car.

When we reached the hospital, a nurse led us down seasick-green halls to a small pink labor room. Pink—the color of accommodation, I thought. Alex was not accommodated; he was dismissed to stand in the hall while I put on a hospital gown.

But it was a great relief to me to have Alex by my side, steadfast and true, during the delivery of our child. He fed me dabs of honey to keep my blood sugar up. He massaged my feet and tried to entertain me. When labor still didn't start, he folded his large frame into the confines of a molded plastic chair, and tried to snooze while I fell asleep.

When morning came, my doctor insisted on taking an X ray to check the baby's position. The X ray showed that our child was in the right position for a natural delivery. The doctor hooked me up to an oxytocin drip, and contractions began immediately. A monitor was also strapped around my belly. Because it kept slipping off, the nurse insisted that I lie on one side to keep it in place. It seemed ridiculous to me that I should be in this uncomfortable position for the sake of technology. If Alex hadn't been there to divert me, I would have cast it aside. This didn't feel like natural childbirth to me.

Just after noon, our daughter was born. She was a darling girl with rosebud lips and a hairline that made her look like Napoleon when she scowled. We named her Anne, after my mother. A nurse immediately took her away to clean her up. I wanted to hold Anne, but she was in another room, being measured, tested, and footprinted.

Once I was back in my hospital room, Alex took his leave to go home and get some rest. "Could I please have my baby?" I asked the nurse.

"You can't have her until feeding time, at four o'clock," she said. That was three hours away.

"May I have something to eat?" I asked. I hadn't eaten for twenty-four hours.

"No," the nurse said. "You just missed lunch. Dinner will be served at six."

No baby; no lunch. I was wide awake and ravenous. By the time I found an aide who agreed to get me a sandwich, all I wanted to do was find my baby and go home. That wasn't possible, so I ate the sandwich and eventually took a nap. When Anne was wheeled into my room in a plastic box, I woke up, looked at her, and, being disoriented, thought, Who are you?

On his way back to visit us that afternoon, Alex went to a "granny trap" in Southampton and bought a daffodil yellow outfit and a soft blanket for Anne. He also bought a split of Château d'Yquem, the famous sweet wine from Sauternes, for the two of us. In the womb Anne had always become especially active the few times I had consumed Sauternes, so he thought it fitting for us to toast each other with d'Yquem, the world's finest. He let Anne suck a drop of it from his finger when she was just a few hours old, as a gesture of bonding, and of appreciation for the natural miracles that make life sweet. After that, we didn't care what happened to Richard Nixon.

The best thing about the newborn baby was the sweet smell of her soft head. The second best thing, for me, was holding her warm body against my breast as she nursed in a deliberate but dreamy way. As soon as I felt the milk flowing, I was overcome with an enormous feeling of relaxation, and my mind was free to wander. I thought about how even though Alex and I had planned for this child, she was not exclusively ours. She was her own little person, with ears that looked like mine and eyebrows that looked like her father's. I thought about how silly I had been to worry about my small breasts, which worked just fine. And I thought back to how we had gotten into this farming life, this vineyard, in the first place.

3

Restless Beginnings

THE SAGA OF THE VINEYARD BEGAN WHEN I MET ALEX. IT WAS only eight weeks after I started my freshman year in college, in the autumn of 1965. I had enrolled at Smith, a women's college in Northampton, Massachusetts, after spending three years at Concord Academy, a girl's boarding school just outside Boston. Concord was a wonderful place for me—a classic New England town and a small school full of interesting people, with plenty of humor and intellectual excitement. It was fine that there weren't any boys there—we had dances with boys' schools to socialize us, if you can call it that—but by the time I graduated, I was ready to be in the company of men. I knew it, too, but I went on to Smith because it was a leading women's school at a time when most of the top colleges that enrolled men were not open to women. My college counselor wouldn't let me apply to Radcliffe, then the women's division of Harvard—she was recommending other girls, with higher scores, for that school.

I wanted to bust out of Smith from the first week I got there. Northampton was, at the time, a seedy old town in the middle of nowhere—at least, that's how it seemed in comparison with Concord. It wasn't the fault of Smith that I was bored; it could have been running a circus and I would have wanted to escape,

In those first weeks as a freshman, I went to a few mixers and met a few pawing guys from nearby Amherst College. After that, it was with great enthusiasm that I accepted an invitation from an old friend of mine—a smart girl from New York who had made it to Radcliffe—to attend a party she and her boyfriend were giving near Boston.

The party was the weekend of the Harvard-Princeton football game. I was already going to be there to meet up with my parents, who thought the game was a good way to get together with me and my older sister, Wendy, who was in nursing school in Boston. My younger brother, Evan, came along, too. We all rooted for Princeton because my dad had gone there, as had a long line of men on both sides of my family, going back to 1832, when a Welsh ancestor of ours had walked from the orchards of Pennsylvania to attend Princeton's theological school.

After the game (Harvard won) and dinner with my family, I got a ride to the party at a house in Cambridge. My friend assured me that there would be plenty of other people without dates, but for the first two hours I found myself alone in a roomful of strangers, dressed uncomfortably in a little black velvet sheath, the only girl without a date. Holding a glass of whiskey and chain-smoking in order to look older and more alluring, I was waiting for Prince Charming. At the time, it wouldn't have occurred to me that Daniel Boone would have been more useful. Buying into the all-American myth of the time, I always believed I would fall in love at first sight and marry a prince. I hoped that I would be discovered for my sensitivity. I wanted someone to take care of me, to provide for me, and I was willing to work very hard for this protection. I tried to please, being aware of the other lessons offered by fairy tales, that there are tests in life, and secrets to unlock—usually secrets controlled by ugly, scary monsters who have to be tricked or kissed.

If a princess can be identified by her sensitivity, how can one tell who is really a prince? I was sure I would know him when I saw him, and I did. The door to the party opened and in he walked. He was a Princeton student on a road trip, along with his roommate, who had a girlfriend in tow. His name, Alexander MacKenzie Hargrave, sounded aristocratic. He was six and a half feet tall—a foot taller than I was—and very slender, like me. He had muscular, Popeye arms, a high brow, chestnut hair a little lighter than mine, and chameleon eyes. His skin was very pale, smooth, nearly hairless. His lips were full—he pulled his lower lip all the time and gave it a little twist. Enchantment was in his soft gaze, a mirror to his mind, turning fierce if necessary. It bespoke power, control, humor. A smile was ready but secondary to his wit. His hands were huge, and I noticed that he walked with a giant's stride. He and I started talking and didn't stop until the party ended. Not being sure of the ways of the world myself, it thrilled me that Alex could speak his mind clearly. What I heard him say was full of surprises, original and smart.

I felt I had to be a little proactive when I met my prince, because first I had to break a curse. The wicked witch was, in this case, my uncle Johnny, who had knelt down to whisper in my ear on my fifth Christmas, just after I got eyeglasses. "You know," he said, "men seldom make passes at girls who wear glasses."

I had to wear my glasses most of the time because everything was blurry without them, but if I went to a party, I took them off. From the time I was in my early teens, my mother, who was beautiful and loved parties, sent me to dances where I didn't know the boys, and I didn't know what to say to them. To solve the problem on the rare occasions when they asked me to dance, I would dance very close and then, a few times, I'd make out with them. That was easier than talking, and it was perfectly safe, right on the dance floor; but the chaperones didn't like it, and in high school I got banned from mixers.

I thought it was an insurmountable obstacle, too, that I was flat-chested. I had to bear the humiliation of wearing padded bras. Buying them was as traumatic as wearing them. The solicitous matron, knowingly nodding at my family's tragedy—having a girl with tiny breasts—would bring out increasingly denser foam to simulate what I lacked. Although I was as girlish as a girl could be, shyly hiding behind my mother and clutching my doll while my sister and cousins climbed trees and rode ponies, when I reached adolescence I couldn't even get a female role in my all-girls school play. Once, I was cast as a boy—a prince, actually—and once I was a bug. I refused to act in school plays after that.

Even without my glasses I could appreciate Alex's finer points on first sight. Two days after I met him, I ran into my best friend at Smith between classes. Standing in the stairwell, I took her aside. My face was flushed and I was still hopping with excitement. "I met someone over the weekend," I said. "I have never met anyone like him before, and I don't think I ever will."

I told my friend how handsome Alex was and how much we had in common. We had learned a great deal about each other in that one evening of conversation. Both of us were New Yorkers, although he had grown up upstate, while I was from Long Island. Alex's father had gone to Princeton, like mine, and while they hadn't known each other there, they had had many friends and experiences in common. And like all four of our parents, both of us had gone to boarding schools in New England. Our families were Christmas and Easter Episcopalians; we liked some of the same hymns, but neither of us could sing worth beans. We had both traveled in France with the Experiment in International Living, a student travel group; we loved French food and could converse, haltingly, in French. It didn't surprise me that we also shared a love of reading and talking about books. What I didn't describe to my friend was the tremendous field of energy that I felt around Alex. This, more than anything I said about him, was what attracted me to him.

Alex invited me to visit him in Princeton a couple of weekends after the party in Cambridge. He was as suave and debonair as I had remembered. We kept on dating each other almost exclusively until we were married almost three years later in 1968. During my sophomore and junior years of college, I spent nearly as much time at Princeton, hanging out with Alex and his friends, as I did at Smith. I made a point of taking courses that met on Tuesday and Thursday (instead of Monday, Wednesday, and Friday), so that I could have long weekends with Alex. My parents loved Alex, too, and as long as my grades were good, they thought it was fine for me to visit him. My mother hadn't gone to college at all—World War II got in the way of that, and besides, she married my father when she was eighteen. I think she was happy to see her shy daughter having a social life at last.

I wasn't wearing glasses when I met Alex, but I put them on the second time I saw him, thinking that if he was going to be my one and only, I'd better get the glasses test over right away. It didn't seem to bother him at all. Then, there was the flat-chest issue. I met Alex's mother and one of his sisters shortly after meeting him, and when I found out that they were slender, like me, and his favorite grandmother was flat as a pancake, I stopped worrying about it. I decided that he was genetically preconditioned to accept flat-chested women.

I was glad I met Alex when I did because I was about to give up on waiting for a prince. Chances are that I would have had sex with the next guy who came around, so I was glad that when it finally happened it was with Alex. We led up to it slowly, spending several nights together cuddling, like the old Puritan custom of bundling. In the middle of the night, we would get the giggles. The slightest thing would set us off laughing. It was like being at a sleepover party with best friends.

Alex proposed to me one night, at the end of a college party, after over two years of steady dating. We had had enough of college weekends and were ready to move on to the next stage. A few months later, after he graduated from Princeton, we got married. Having fulfilled the requirements for my major in government at Smith, I was able to take my last year of college at Harvard while he studied for his master's degree in Asian Studies. While at Princeton he had decided not to become a lawyer like his father and grandfather. He chose to join the small group of students that were studying Chinese—the first step he took away from doing what his parents expected of him. The challenge—and the inscrutability—of Chinese studies so suited him that he decided to pursue it into graduate school.

Alex taught me two phrases in Chinese. One sounded like *"How bu how"* ("How are you?"); the other sounded like *"Bu juan lien, bu juan lichee,"* a phrase from Chairman Mao that means "If you don't exercise, you won't build up strength." Little did I know how pertinent that phrase would turn out to be.

Alex loved Chinese poetry, but his college thesis had been about politics. He had researched Mao's idea of perpetual revolution—the deliberate creation of constant change—and learned that it had its roots deep in Chinese culture. He had also read that in our own country, back in the eighteenth century, Thomas Jefferson had had a similar idea, that the ownership of land should constantly change and land should belong only to those who actively farmed it. That seed of an idea would grow over the next few years, until we became farmers ourselves.

Those were the days, during the Vietnam War, when a newly-wed couple in school didn't necessarily have to plan a future career. It was enough to be enrolled in school. Alex would not have been drafted by the military anyway because he had a spinal curvature—probably the result of exposure to polio as a child—but many of our friends in school were relieved to have student exemptions from the draft. Instead of talking about our careers, we talked about ideas. All of us wanted the war to end, and the air was charged with angry protests. A student strike at Harvard that shut down classes just before final exams nearly kept us from finishing our degrees. There was a strange mixture of fervent political unrest and equally fervent partying that went on that year at Harvard. Frisbees flew in the Yard as students made silkscreen T-shirts with peace signs and big red fists. We were opposed to the war, and we participated in some of the protests, but for the most part, we focused on making our fifth-floor walk-up apartment a domestic haven from all the turmoil.

On a Saturday in late September, I made a foray to the local supermarket, which was at the bottom of a long hill. I wanted to be a good housewife, so I thought I should look for bargains. My prize was a long cylinder of baloney, five pounds for a couple of dollars. Resting it against my shoulder like a baguette or a baby as I carried it upstairs, I thought of all the thrifty meals I could make with it. Alex met me at the door. He saw the chunk of baloney and declared that he never touched the stuff. An iridescent wedge of it remained in our refrigerator after I ate about three pounds of it myself. It weighed on my conscience. A few weeks later, I quit smoking and threw out the baloney with my cigarettes.

That was in November, when I also decided to give up bras. On that day, I was wearing a special one my mother had bought me for my trousseau. It wasn't my wedding bra, which had white lace and a tiny blue satin bow in front, but a Maidenform with polyester padding. It had a diffuse orange floral pattern on it, as if it were trying to match a hippie aesthetic, only hippies don't wear bras. I wasn't a hippie, but I quit wearing a bra when other women were burning theirs. No need for a fire, I thought. I'll just leave it in the drawer.

That liberating gesture came at the same time I figured out that I didn't need to go to class very often in order to pass my courses. I had signed up for American social history and several social relations courses, also known as sociology—the easiest gut courses Harvard had to offer. After attending one such course for a couple of weeks, I got tired of hearing the professor reading from his own book while bits of spittle collected on his lips and were finally projected into the ever-diminishing audience. While I continued to do the required work for the course, I checked into that class only periodically. This was the first time in my life that I hadn't devoted myself to my studies, but my grades were good anyway, and I thought it was more important to devote myself to my new husband. With time on my hands, I dove into the cookbooks we had been given as wedding presents. Rummaging through the pages of *Mastering the Art of French Cooking*, by Julia Child, I found my solace and delight. Besides, wasn't food the way to a man's heart?

I had naked ducks hanging by a string above the kitchen counter, drying for Peking duck. I cooked daubes and soufflés, quiches and tarts. Alex helped out, often choosing unusual things for us to try. Once, he found himself in line ahead of Julia Child herself at the butcher. When he ordered a rabbit, cut up, her voice boomed out behind him, saying "Be sure to cook the head!"

Our student apartment was dingy and unfurnished, with a barely functional kitchenette and a view of a brick wall. My grandfather gave us a bed as a wedding present, saying "This will be the most important thing in your marriage," and Alex's grandmother gave us some money to buy furniture. In Boston, down on Charles Street, there was a row of antique stores we frequented. We bought a red-lacquered Chinese throne with gilded designs of bucolic scenes. We bought a portrait of a child in a blue empire-style dress, circa 1820. It was done in pastels, and the background was smudged. Then we bought a carved Chinese wooden funeral urn, shaped like the head and torso of the man it memorialized. Our friend Wheeler, who was studying Sanskrit and Arabic, told us that we had to open the back of it, where we could see a resin-sealed little door. He was afraid there were still ashes there that could bring bad luck. There might be something valuable in there, too. He offered to do an exorcism for us, so we cooked him a big meal, and then, after dinner, we lit candles around the wooden figure. We knelt beside it, chipping away the resin until we could pry open the back. "Allah, Alla-ha, Allah," Wheeler intoned. The panel opened, but the urn was empty.

Our friendship with Wheeler led us into some partying, which ended with our deciding not to drink distilled spirits anymore. As college students, we hardly ever drank wine. We drank bourbon, as did our friends. Well, some of them drank beer, but I disliked the bitter taste of hops. To be honest, I didn't like the taste of bourbon much, either, but it was better than scotch, which smelled like tin cans, or gin, which gave me a headache.

One Sunday morning, after an evening of partying, we heard loud knocking at our door. Alex was flat out, and I was barely conscious. I stumbled into the hall and started to open the door. By then Alex was awake. "Don't open the door!" he shouted. It wouldn't have occurred to me not to open it, but I let go of the handle and asked the visitor what he wanted. "Something happened to my friend and I need to use your phone," he said. I was about to let him in again when Alex told me to wait. He called the police, who said we should keep talking to the man until they came. The man kept saying "Let me in; it's urgent, it's urgent." Within a couple of minutes, the cops were on our landing. They handcuffed the intruder, who they said was wanted for multiple rapes in the neighborhood.

I thought about how close I had come to opening the door, and I blamed my carelessness on my hangover. I decided that I didn't want to be impaired by alcohol. Alex and I had a long discussion about it. We talked about how our parents usually had a couple of drinks before dinner and how we were starting to do the same. Alex pointed out that wine is the beverage of civilization, unlike whiskey. We agreed to give up hard liquor but not wine, which we would drink if we had friends over for a meal.

When we visited Alex's family in Rochester that Thanksgiving, we were warmly greeted by all his relatives at his grandmother's house for the traditional turkey dinner. When we were offered drinks and asked for ginger ale, our reception became distinctly cooler. My relatives treated us the same way at Christmas. It was astonishing how upset our families were that we didn't want to join them in a cocktail. We didn't make a big deal of it, but they took it as an insult that we were drinking ginger ale or wine while they were drinking gin and whiskey. We wanted to be less like them after that. We got more interested in wine, not just because it was another form of alcohol, but because it was something we were starting to know more about than our families. I liked all the choices offered by wine, and the search for my own preferences.

Like other students, we began drinking Lancer's Rosé and Liebfraumilch. Soon, we discovered that there were much better wines to be had, still for under four dollars a bottle. That's when we got into drinking vintage wines from Bordeaux and Burgundy—wines that are made from Cabernet Sauvignon and Sauvignon Blanc, Pinot Noir and Chardonnay, which became the first grape varieties that we eventually planted.

Wine appealed to me for the way it engendered relaxed conversation. Beyond that, it had a deeper appeal. In holding a glass of red wine, swirling the glinting prism of reds, pinks, and purples, I returned in my mind to my maternal grandmother's house. Hints of sandalwood or cedar, or the aromas of white flowers, in wine reminded me of the distinctive scent that permeated her house. In her guest room, she had an ancient cabinet that I liked to open to smell the strange, musky aromas inside its drawers, like the aromas of a very old wine.

My grandmother was an artist who traveled often and spent part of each year in Mexico. There were bundles of multicolored woolen Mexican ribbons hanging from the mirrored doors of her closet. She let my sister, Wendy, and me each choose a color from the bundles to tie back our wet hair after we went swimming with her. While we made our choices, we would play with the mirrored doors, opening them both while we stood between them so that our reflections, and the reflections of the ribbons, were repeated infinitely. This, too, was like gazing into the depths of a glass of wine, where I could see my image, or the light from a candle, or the shape of a companion, inside the whirlpool.

The experience of a great wine can be disquieting, too, like my grandmother's art. Wine's superficial allure sometimes has an undercurrent of disagreeable bitterness—things that challenge appreciation and keep you wondering if you like it or not. Granny's paintings showed beautiful scenes from her travels— Venetian domes, African zebras, Maine by moonlight—which gave me a feeling of peace, and just a hint of foreboding. Time was suspended in them, as if she knew this world wouldn't stay the same. She died at the age of sixty-one, when I was thirteen.

***

We were just barely beginning our involvement in wine, and hadn't come close to thinking of it as a career, when the school year ended. Our lives took a major detour when the spinal curvature that had kept Alex from being drafted suddenly deteriorated. It required surgery for multiple fusions. We moved to his home town of Rochester for the operation. His parents rented us a cottage just down the street from their house. I worked in our landlady's nursery school in the mornings, but essentially I became Alex's nurse for the next eleven months, while he remained on his back in a body cast.

During the seven hours that Alex was in surgery, I was alone in our cottage with nothing to do. I started arranging furniture and moving objects around the living room where Alex would be staying after his surgery in a big hospital bed. The lacquered Chinese throne that we had bought on Charles Street was looking dingy, so I decided to clean it. With a bottle of rubbing alcohol, I rubbed the front splat vigorously. Thinking of my husband undergoing radical surgery, I failed to notice that I had rubbed the gold leaf design right off, along with the dirt. When I finally saw what I had done, I felt that in my carelessness I had damaged more than the chair. The chair didn't matter; I was worried about how to be a good wife to someone who was in pain and encased in plaster.

The next months were indeed a time of trial for us. I had married Alex because he was someone I could depend on, and now he depended on me. His parents came by periodically to offer support or to play a game of cards, but I was essentially as confined as Alex was. I didn't want to leave him alone for more than an hour or two, and besides, for most of that year I had to borrow a car if I wanted to go anywhere, and I had nowhere to go. We could still enjoy our meals together; we both tried to do a little artwork, making Christmas decorations and a couple of angular acrylic paintings. Alex wrote elaborate poetry and read *Worlds in Collision* by Immanuel Velikovsky.

While it wasn't anything either of us would have chosen, as the year went on, I became a pretty good nurse, and Alex was a good patient. We became so used to each other's constant company that by the time Alex was released from his cast, we were thoroughly bonded, but more like siblings than mates. It was time to reignite the flame. Spring came with all the usual signs of rebirth. The stream next to our cottage was swollen and gushing. It was a liquid highway for ducks that quacked gleefully as they rode the waves, mating as they floated past our window.

The instinct to flee did not assert itself right away. Having taken this year for the surgery, now we needed to get back on track like any other young couple, but we didn't know what to do. Alex decided that he had gone as far as he could go with his Asian studies. He told me that to go back to school for a doctorate, he would need an Asian wife to translate his work for him. He had already learned Mandarin and classical Chinese, and would have to learn Japanese to go on.

Less than forty-eight hours after Alex had submitted his thesis to his college professor, the CIA had tried to recruit him. He had been appalled that the route from his professor's study to the CIA was so direct, and anyway, being six-six would make it hard for him to sneak around China. The summer before we were married, he had been in Taiwan and Hong Kong, and little old ladies had given him their seats on the bus because he was too tall to stand up.

A solution to our quandary presented itself when my father found that one of his colleagues in publishing was looking for a trade book editor in Boston. We liked Boston, so back we went. I enrolled at the last minute in a master-of-teaching program at Simmons College, a logical place for me after my year of teaching nursery school in Rochester.

What seemed at first like a good fit turned out to be less than ideal. Alex could not sit in a chair, confined to an office for eight hours a day. His spine had healed, but he had not had any therapy to rebuild his muscles. Weighing a slim 165 pounds before surgery, he emerged from his cast like a turtle missing its shell at a meager 135. The walks to his office across Boston Common and back to our apartment were the best parts of his day. He needed to build his strength, and we both still needed to somehow free ourselves from what felt to me like the giant conveyor belt of conventional life.

On the pretext of trying to get admitted to China via the new Chinese consulate in Toronto, using a contact with the Dutch consulate—an old friend of my grandparents—Alex quit his job at the same time that I graduated from Simmons in May 1971. While we managed to gain an interview with a group of Red Guards who were living under heavy security in an apartment in Toronto, the trip to China did not pan out. Had we known more about the Chinese Cultural Revolution, we would never have imagined that they would welcome two young Americans who carried backpacks, one of whom spoke Chinese. This was almost a year before Nixon visited China and reestablished American relations there.

From the Chinese consulate, we went on by bus to visit my cousins in Maine. A friend lent us her summer cottage on Deer Isle for a couple of weeks, and it was there that we fell upon the books of Adele Davis. She was a popular nutritionist who sang the praises of whole grains, dairy products, and organ meats. Her books—*Let's Eat Right to Keep Fit* and *Let's Get Well*—appeared like an answer from heaven. They would tell us how to regain our health.

Not having a car, we put on our backpacks and hiked for a couple of hours to the only grocery store on Deer Isle. There, we loaded up on pork liver—something that Adele urged us to eat in great quantities. By the time we got back to the cottage, Alex had raging blisters from his new hiking boots, and a fierce allergy attack from all the pine trees, but we were full of excitement about our new healing diet. The excitement lasted until we actually ate—or tried to eat—the pork liver. Surely, we concluded, that was not food meant for human consumption.

We tossed the rest of the liver and started to focus on gathering wild food, following the instructions of another natural food guru, Euell Gibbons. His book *Stalking the Blue-Eyed Scallop* was nestled alongside Adele Davis's volumes, and we were in the perfect place to take advantage of it. We made omelets out of tiny periwinkle snails that we plucked from their shells one by one with a straight pin. We shucked minuscule beach peas, and gathered salty grasses that looked like green beans. Hungry— really hungry—we went back to the grocery store for more organ meat and found that beef heart made an edible steak.

Sometime in the midst of this wild retreat Alex had the idea that if organs were so nutritious, it would be worthwhile to develop techniques for their preparation that would make them taste good. This was a challenge that was most appealing— something that we could do together, and something that no one else had done. We would be the gurus of organ meats.

In order to work on our cookbook we spent a few months living at his grandmother's cottage on Canandaigua Lake, up north in New York's Finger Lakes district. We were surrounded by vineyards. The beauty of the vines started to have an effect on us, and we began to muse about a life as vintners. It was a highly romantic notion—a way for us to work and raise a family together on our own terms. We always wanted to have children. We also wanted our children to be near us, so that they would know where we were and what we were doing. When I was a child, it had bothered me that my father disappeared to go to work at his editing job in Manhattan each morning before I was even awake. My mother stayed at home, but where was my father? What was this "work" that he had to do that was so important that he had to leave us for the whole day? He carried bundles of papers that he read even when he was at home. Alex, whose father also disappeared to an office every day, agreed that for him, too, father=absent=work. We thought that the equation for farming would be family=home=work. Although I was a woman and therefore expected to stay at home, I wanted to have a meaningful job in addition to raising children. If my children could see what I was doing, then maybe work would have meaning for them, too. Our work wouldn't produce a sheaf of papers; it would produce fruit that we could touch, taste, smell, and ultimately preserve as wine.

When we drove around Canandaigua Lake, we often passed a dairy farm, which was at the top of a hill with a view of all the lake. When he glanced at it, Alex's face looked pained as he described to me how the farm used to belong to his grandfather back in the twenties, until he lost it after the stock market crashed in 1929.

His grandfather, T. J. Hargrave, came from a Quaker family in Nebraska. He had been recruited by the head of Harvard Law School and graduated with a law degree at the top of his class. When he joined his roommate to found a law partnership in Rochester, New York, he brought his midwestern love of the land along. He bought the Canandaigua farm with money he borrowed on the basis of stock he'd gotten from his work for George Eastman (the inventor of the Kodak camera). When the market crashed, T. J. couldn't cover his loans, and he went bankrupt, losing the farm. In 1932, Eastman shot and killed himself, leaving a note saying that he wanted T. J. Hargrave to run Eastman Kodak. Even though T. J. left his law firm and ran Kodak for thirty years, he never got his farm back.

None of the Hargraves could drive by that dairy farm without sighing, or talking about how T. J. had loved the vegetable garden in his backyard, where he had tilled the soil every evening after work through the war years. There must have been something in his grandfather's legacy, a longing for the land, that affected Alex, too.

Although we didn't attend any church regularly, Alex loved to quote from the Bible about sitting "under his vine and under his fig tree." The idea of a peaceful farm life meshed with his college readings in Chinese poetry and philosophy—an approach that exalted meditation and regarded farming as the ultimate meditation, besides being a confirmation of one's right to use the land.

Near where we were staying on Canandaigua Lake, we rediscovered a childhood friend of Alex's, John Ingle. Alex and John had been urban Cub Scouts in the same den when they were growing up in Rochester. They had gotten badges for standing on the roof of a building in downtown Rochester and identifying cars as they went by. On the shores of Canandaigua Lake, John and his wife, Joey, had an organic farm and vineyard that looked idyllic to me, even though there was no running water in their cabin. It was cold there, though, and we found out that they were having problems growing the kinds of grapes that made the kinds of wines we liked, the *Vitis vinifera*. There the grape varieties that make the best wine would freeze back to the ground almost every other winter when temperatures fell below zero.

Just how cold it was impressed me forcibly when we volunteered to help harvest grapes at another neighboring vineyard, where the typical upstate Concord grapes were grown. We showed up in the middle of an early October morning, after a crew of seasoned harvesters had made a good start on the picking. The vineyard manager showed us how to loosen any clusters that were tangled up with shoots, and gave us needle-nose clippers to snip each cluster at its stem. It started to snow shortly after we began to pick, and then the wind started up. It was much harder than it looked to pick the clusters without losing fruit to the ground. Not only was it frustrating; it was also boring. We quit when our hands got cold—maybe fifteen minutes after we started. Some of the others quit at the same time, but part of the crew— the ones who were being paid—kept on going. After we got home, I felt embarrassed. I had wimped out, and I did not want to be a wimp.

When I was a little girl, and throughout my childhood, I had been a bookworm who couldn't quite keep up with my very athletic older sister, Wendy. When I was about six and Wendy was seven, we used to play cowboys and Indians. Wendy had a confident swagger that went with her Annie Oakley costume. She had a red cowboy hat with fancy braided trim like a lariat circling the brim, and a fringed vest that she wore over her smocked dress. With her official Annie Oakley holster and cap pistol, she looked tough despite the way her puffed sleeves stuck out. I didn't have a cowgirl outfit, so I had to play the Indian. We both knew that Indians never win. Wendy would chase me around the house, firing her toy gun, and eventually I would fall to the ground, play dead for a few seconds, and then we'd start all over again.

Evan was four years younger than I was, but even he ran circles around me once he got to be old enough to play in our family softball or touch football games. I was always afraid that the ball would hit my glasses, and by shying away from the ball I became less and less of a participant in athletic events and more of a girl who stayed inside. Now, in the vineyards of Canandaigua, I wanted to change all that. Still, I came up against my own lack of stamina.

One of the things that had made me love books also had a deep effect on my desire to become a stronger, more independent girl. My paternal grandfather started to read me the Little House books, by Laura Ingalls Wilder, at about the same time that Wendy and I played cowboys and Indians. When he was off traveling, I grew frustrated waiting for him to come back and read me the next chapter, so I learned to read the series of books myself. I wanted to be like Laura, the pioneer girl who survived several volumes' worth of hardship in the American West of the 1870s. The period Wilder described happened just about ten years before my grandfather was born; and even though her books were written for children, they conveyed a way of life and a state of mind that was part of his experience, too. Everyone's life then depended on cooperation. Small details—an orange at Christmastime, a dance to the fiddle, a meeting of friends— were greatly appreciated, and nothing was taken for granted. Laura, like me, had a blond-haired older sister who was always perfect, whereas we were dark-haired and made mistakes. Laura was physical and up to any hardship. I dreamed of having a chance to build my strength and test myself with hardship.

The same grandfather who read to me about the western pioneers affected me with his own life story, too. He was known to us as Big Dad, but his real name was Norman Thomas. He told me stories of what it was like for him to grow up on the edge of the frontier, in Marion, Ohio. He was the son of a Presbyterian minister and could recite any part of the Bible, although he had rejected formal religion in favor of Socialism. In fact, by the time I came along, he was a prominent Socialist who had run for president of the United States five times. That wasn't what interested me about him when I was growing up; no one could adequately explain to me what these political views meant anyway. I did understand that communism was bad—we had air drills in school because we thought the Commies would bomb us—and I knew that Big Dad had warned against communism from the time of the Russian Revolution. It impressed me that he seemed to remember events and people in great detail. In fact, just a couple of years before his death in 1968, when he first met Alex, Big Dad had looked at him carefully and said, "Ah, yes, Hargrave. I met your grandfather [T. J. Hargrave] at Canandaigua Lake in the thirties. I was making a speech in town, and he came to hear me. When my speech was over, he came up to me and said, 'Mr. Thomas, I don't agree with anything you say, but I like you. Won't you come to my house for dinner?' "

Big Dad had gone to the house—the same house where Alex and I first thought of growing grapes—and had a cordial debate with the capitalist from Nebraska.

When I was little, what impressed me about Big Dad, besides his childhood stories, was the deep caring he expressed for people, for ideas, for anything that was important to him. It wasn't enough to have ideas; he felt it was essential to act on them, to get things done, not to wait or expect anyone else to do them.

I was also touched by stories he told me about his mother and her parents, the Mattoons. They had been missionaries to Siam (now Thailand) before the American Civil War, and translated the Bible into Siamese. Anna (of *Anna and the King*) later replaced them after they returned to the United States for lack of funding. Big Dad's mother thus grew up in the court of the king of Siam. She lived there until she was eight years old. It was this experience of his mother, and his own trip to China shortly after his graduation from Princeton in 1903, that gave him a deep sensitivity to the plights of other people.

I sometimes feel guilty that I was not a social crusader like Big Dad. There must have been a touch of his zeal in the way I threw myself into growing grapes—but that wasn't for more than another year.

⁂

After our experience picking grapes in Canandaigua, while the glimmer of an idea to grow grapes was still there, we had our organ-meat cookbook to work on. I had a degree in teaching, and that was something I had always thought of doing. In the fifties, when a little girl was asked what she wanted to be when she grew up, there were only three acceptable answers: a nurse, a secretary, or a teacher. Wendy wanted to be a nurse, and I liked school, so I picked teacher. When I was in high school, I identified with my English teacher, Miss Stone, because, like me, she was trying—unsuccessfully—to get used to contact lenses. She also looked a bit like me, with a slight figure, even features, a pointy nose, and curly brown hair. Miss Stone was right out of

Smith College and new to teaching. She was bright and pleasant, but my classmates and I saw her lack of confidence as license to create havoc in the classroom. I think we gave her a nervous breakdown.

One of the books we read under her tutelage was *My Antonia*, Willa Cather's book about a young immigrant woman growing up on the prairie in the 1870s. Here was another heroine like Laura of the Little House books, a strong pioneer woman who gloried in doing physical work but was always cheerful regardless of hardship. How could I continue to identify with Miss Stone when I read this about Antonia? "She had only to stand in the orchard, to put her hand on a little crab tree and look up at the apples, to make you feel the goodness of planting and tending and harvesting at last. All the strong things of her heart came out in her body, that had been so tireless in serving generous emotions."

Maybe being like Antonia would suit me better than being like Miss Stone. For the time being it was a moot point, since I was committed to working on our cookbook. Alex and I went to a slaughterhouse in Rochester, where we bought every kind of innard we could, including brains and spleen. We spent months researching the nutritional benefits of organ meats, which are loaded with vitamins, and the history of "pure foods." Recognizing that the biggest obstacle to getting people to eat organs was that unlike muscle meats they are identifiable as body parts, we investigated the history of cannibalism as well. Apparently, when the U.S. government sent canned baby food to New Guinea after World War II, the natives wouldn't use it. The cans had pictures of babies on them, and the natives thought that the cans contained puree of baby.

At the same time we researched background material for the book, we invented all sorts of dishes using the meats. It was actually a great deal of fun, and we came up with some good techniques for cooking the organs. When the book was done, we even got an agent to represent us in trying to sell it. Being in the book business, my father had asked one of the agents he dealt with to represent us as a favor. Come to think of it, this poor agent did have quite a distressed look on his face when we presented the book to him—but he gamely sent the manuscript around.

In evaluating the worthiness of our book we should have paid more attention to the response of some old grade-school friends of Alex's who attended a dinner we gave to feature our new cuisine. They gnawed a bit on the beef heart teriyaki we served for hors d'oeuvres. At the dinner table, they politely admired the lamb's testicle that we had cooked and decorated to look like a chicken, with a parsnip beak and chrysanthemum petals for feathers. There was no way they were going to eat that thing once they found out what it really was. Fortunately, we had served some very good wine with the meal—a Château Beychevelle or a Beaune-Villages. By the time we got to serving the tortoni-style ice cream that was made out of whipped calves' brains, whipped cream, and sugar, even though our guests hadn't eaten much more than the bread on the table, they were all drunk enough to congratulate us on our cookbook and wish us well.

4

Westward Ho! and Back

WE HADN'T EXACTLY MADE UP OUR MINDS TO BE WINE PIO-
neers yet, but in the spring of 1972 we did heed the wanderer's
call to go west. Not having jobs, we were house-sitting in order
to reduce our expenses while we waited to sell our cookbook. A
friend of ours invited us to stay in his family's house in Monterey
while they made a tour of the Middle East for several months.
Vaguely thinking that we might find a vineyard, or go on to
Hong Kong, or just see what else developed, we bought a used
Jeep Wagoneer, fitted it with a foam mattress, packed three
changes of clothes—which included a nylon polka-dot dress for
me—and headed out to the West Coast.

In northern Colorado we took a side trip into Dinosaur National Monument, armed with a Texaco map that showed a nice
little trail called the Yampa Bench Road, a dirt track that mean-
ders for forty miles through the canyon lands. Thinking that
this would be a pleasant detour, we left the campground where
we had spent the night and followed the map to the trail.

The road was paved for about twenty miles, and we congratu-
lated ourselves on having decided to go away from the tourist
area into this beautiful open land with its interesting rock for-
mations. We stopped the car to take photos of prairie dogs and
Indian hieroglyphs, and we ate a picnic lunch, sitting on a

smooth granite rock overlooking the beginning of a deep canyon. Shortly after we got back on the road, it narrowed until it became just the width of a vehicle, with canyon cliffs on both sides and no guard rails. That's the part of the road that was called a "bench," we realized.

Just before we got to the narrowest section, it started to snow. The soils there are full of bentonite, ironically enough a fossil that is used to clarify wine in winemaking. When bentonite is wet, it has the consistency of buttercream frosting—not a great surface to drive on when there are cliffs on either side of the road. We couldn't turn the car around because the road was too narrow, so we kept on driving. It would have been safer to just stop for a while, but we had eaten our last bit of food for lunch. We thought the storm might be lengthy, and we didn't believe that anyone would find us for a long time if we got snowed in. No one knew we were there. It was early spring, way before the tourist season would bring other visitors to the area. Besides, we thought we were too young to die.

We got over the bench, only to find that the only way out of the park entailed a drive up over a nearly vertical ridge, on switchbacks that made hairpin turns. At this point Alex had been driving for several hours, fully focused on the road. All the blood had drained from his hands and face. As he revved the engine to make the impossible ascent, a flock of sparrows appeared, hovering over the engine. They seemed to be pulling the car up the hill like a cosmic force. There could easily have been the aroma of roses and celestial music, too, but I only remember the birds. When we got to the top of the ridge, there was a big sign that should have been posted at the start of the trail we had just traversed. It said: ROAD CLOSED—IMPASSABLE IN SNOW. That's when I knew that we were being saved for better things. And from that moment on, I knew that I could trust Alex to guide us if things got tough.

We made it out to California without further incident, other than losing the ten dollars we allowed ourselves to gamble in Las Vegas. When we crossed the border from Nevada into California, we were thoroughly taken aback by the great number of California highway patrolmen who slowed down to check us out as they passed our Jeep. We stopped by the side of the road so that Alex could cut off my long hair. His hair had been short since his operation. We didn't want to be mistaken for hippies.

The Monterey-Carmel area was a haven for hippies in the early seventies. It was a bizarre mixture of quasi police state, movie-star residents, brown rice, and free love. While we might have shared the political views of the hippies who camped out in Big Sur just south of Monterey, we didn't want to be identified with them. It wasn't worth the risk of being hassled by the police. Furthermore, we didn't use drugs. (I have to qualify that by admitting that I tried marijuana once in Boston, but I didn't inhale.) Life seemed vertiginous enough without going through it with an altered mind.

While we were in Monterey, we would sometimes pick up hippie hitchhikers. They never thanked us; I was put off by their attitude that we *owed* them a ride, since we had a car and they didn't. One of the hippies we picked up kept asking us, "What's your odyssey, man?" I found that particularly annoying, even if it was the right question to be asking.

After spending several weeks in California, we learned that our book had been turned down by every publisher that had been kind enough to look at it—all two of them. After our friends returned from the Middle East, we packed up our Jeep again and set off up the West Coast, with a sense of disillusionment that our work on the book had been for naught. We were living on income from the trust Alex's grandfather had left him. I didn't know how much money was in the trust, but whatever it was, I believed it should not be wasted on a western vacation. In my mind, it was okay to live on it as long as our time was spent purposefully. Too many people our age were fighting and dying in Vietnam for us to be frolicking around the country. We were lucky that Alex was exempt from the draft because of his fused spine. Had I been a man, my birth date—September 14—would have made my lottery number in the draft number one, sending me straight off to war. While we spent under three thousand dollars on all of our expenses in 1972, once we heard from our agent that he couldn't sell our cookbook, I felt as if we were wasting our resources. Still, the thought of starting a vineyard was no more than a thought. We needed to find a place where we would feel comfortable living, as well as a place where we could grow vinifera grapes.

It was no challenge to grow and ripen European wine grapes on the West Coast. There was no freezing winter in the valleys along the coast, and the abundant sunshine in many places made it easy to ripen the fruit. The first Spanish settlers who arrived in California in the seventeenth century were monks who planted wine grapes as soon as they landed there and successfully tended them; a secular wine industry was established around them. Although this industry had been gravely hurt by Prohibition, it was being rebuilt by leaps and bounds, and in the early seventies it was starting to boom.

Just east of Monterey, on the other side of a pile of dusty hills, we discovered an entire valley full of broccoli and grapevines. The rows stretched on for miles, standing out in shocking green contrast to the scrubby land around them. There was no sign of human life in this paradise of agriculture—no houses, no villages—just plants. Loudspeakers placed strategically throughout the vines blared out amplified screeching that sounded like a thousand birds being tortured. I guess that was to convince any stray birds that this was not a good place for them to settle. It sure convinced me. I wanted to live in a regular village, with my farm near my house, not out in the middle of some irrigated desert.

We went on to the Napa Valley, north of San Francisco, where we saw gnarled rows of little trees. They didn't look like any grapevines we had seen before. We thought they were prune trees until we realized that they were just shaped differently from the New York vineyards we were familiar with, which are all trellised. These vines had stakes that held them up separately.

We stopped into Beaulieu Vineyard, Heitz Cellars, and Louis M. Martini. Why were they holding wine in redwood tanks? The wines were made from French grape varieties, but they tasted of redwood. The white wines we tasted were sweet and flat on the palate, and the reds had too much alcohol. We must have gone there at the wrong time and tasted the wrong wines. How else could we have dismissed them so quickly? The valley was beautiful, but it was over a hundred degrees when we were there. We were alarmed to see smudge pots throughout the vines, indicating problems with springtime frost. A little like Goldilocks, we barely tasted the region before declaring it both too hot and too cold. Still profoundly unsettled by the year that Alex had spent in a plaster body cast, we weren't sure exactly what we wanted, so it seemed that this was just a preliminary scouting trip. Restlessly, we kept moving up the coast.

We drove north all the way to British Columbia, hiking and camping along the way. Instead of focusing on a site for a vineyard, we sought a level of comfort in our surroundings. Maybe it's not surprising that, finding all these places unfamiliar, we ruled them out summarily. Oregon hills had been raped by loggers. Washington was too wet in the west and too dry and too empty in the east. British Columbians didn't want Americans, and said so.

I liked the camping we did, but not the aimlessness of it. It bothered me when, after hiking for three hours into a wilderness area, we saw a sign that read WELCOME TO THE WILDERNESS.

On that trip we always stayed in national forests because, not having any facilities, they were cheap and sparsely populated. At one of these campgrounds, we pitched our tent in a patch of dense evergreens next to a formidable creek. "The old swimming hole!" I cried enthusiastically and plunged right in. It grabbed me and took me flying past the campground. Scrambling out on the first rock I could grab, I called to Alex to join me. We charged to the top of the creek and repeatedly tumbled back through the water, grabbing onto the rocks. Exhilarated, we clambered together over a hill well beyond the campground where, in a little cave, we lay, barnacled together.

We didn't hear the family of campers coming over the rise until it was too late. Whatever their little boy saw ended his innocence. "I'm sorry, I'm sorry!" I babbled. "We're married!"

If there was no wilderness in the wilderness and no privacy in the middle of nowhere, then it was time for us to find our frontier in more familiar territory. On the way back from Canada, standing beside a lake in Oregon that was full of flamingos whose shocking pink color was intensified by the still waters, I realized that I would never be happy where I felt so thoroughly out of place. I wanted us to act on our idea of growing wine grapes, and I didn't want to stay on the West Coast any longer. Turning to Alex, I threw down the gauntlet. I said, "Let's go back east and plant a vineyard there."

On our way back to New York, where we thought we could put down roots, we bought whatever books we could find about wine grapes. The pickings were slim when it came to books that related to wines in America. Alex and I read aloud to each other as we took turns driving. I chose selections from *McCall's Guide to Wines of the Americas,* a book by William Massee published in 1970 that had basic information about the existing wines and wine regions. Massee's descriptions of centuries of failed attempts to grow vinifera on the East Coast were daunting. A single sentence stood out: "Good wines cannot be made from poor grapes." That was when we knew that it would not be worthwhile to plant grapes unless we could find a place to grow vinifera. We would have to succeed where others had failed. That was exactly the sort of challenge we wanted.

I thought that Alex was just fooling around when he read to me from Virgil's *Georgics* and Columella's *De Re Rustica,* two books that were translations of ancient Roman verses. Nonetheless, they proved to be full of sound advice. Columella and Virgil wrote around the time of Jesus, and offered complete advice for planting and tending grapevines and other crops. They were writing to instruct the Romans who were bringing vines to conquered parts of the Roman Empire. We were astonished by a chapter on choosing the best slaves. It said that criminal slaves were smarter than war-captive slaves because criminals had to think in order to commit crimes, whereas war captives were just in the wrong place at the wrong time. The rest of the texts were inspiring. Who wouldn't be tempted by Virgil, as he described all the aspects of an idyllic farm life:

> Now balance with these Gifts, the fumy Joys,
> Of Wine, attended with eternal Noise . . .
> Easy Quiet, a secure retreat,
> A harmless Life that knows not how to cheat,
> With homebred Plenty the rich Owner bless,
> And rural Pleasures crown his Happiness.
> Unvex'd with Quarrels, undisturbed with Noise,
> The Country King his peaceful Realm enjoys.

When we told our families of our vineyard plans, they took the news in stride. Our mothers were both women who had encouraged us to do whatever interested us, so they could hardly complain now. My father, a man of few but unfailingly honest words, kept his opinion to himself except that he always referred to our vineyard as "the pea patch." I wasn't fazed when my dad's best friend asked Alex, "What will you do with your mind?" We were bound and determined to follow our plan regardless of what anyone thought. It was a good thing that Alex had that inheritance, though, because when he asked his father for financial assistance, his father responded, "Sure, I'll help you. I'll help you by giving you nothing. You have to do it yourself."

While Alex's father didn't want to get involved in a vineyard investment, he generously offered to pay for any further education either of us might want. As we drove back from the West Coast, I decided that having a vineyard wouldn't take much time, so while the vines grew I could go back to school to study science. Alex's father paid for me to enroll in chemistry and calculus courses at the University of Rochester while we scouted for vineyard property.

Although our parents were okay with our plans, it was with some trepidation that we told Alex's maternal grandmother about our unorthodox future. It turned out that she didn't care what we did; she just loved Alex. At a Sunday lunch at her house, sitting at her dining room table and eating crème brûlée for dessert, we told her that we wanted to grow grapes. I remember the crème brûlée because I had never had it before, and its silky sweetness softened my anxiety about discussing our plans. I hadn't eaten more than half a serving when Grandmother's German cook, Rose, who heard our conversation, came out of the kitchen crying. She tearfully begged us not to attempt to grow grapes. Rose described her own family's vineyard on the Rhine, where every year she had carried baskets of eroded soil from the base of the slope back up to the top. Her words were affecting, but I tuned them out and asked for more crème brûlée. I had no plans to grow grapes on a cliff.

The hills around the Finger Lakes, where we began our East Coast search, were daunting enough. It didn't take long for us to discover that cold, muddy soil is terrible for wine grapes. There were a few growers who were trying to grow vinifera there; we saw their newly planted vinifera vines dying of grotesque cancers, called crown galls, that ripped apart their trunks because the soil there was too cold and wet. Virgil was right when he warned the Romans looking for a vineyard site:

Moist earth produces corn and grass, but both
Too rank and too luxuriant in their growth.
Let not my land so large a promise boast.

For a few weeks we were almost convinced that we could make the kind of wine we liked out of some new French-American hybrid grape varieties with names like Seibel 5279 and Baco Noir. They were being touted by plant scientists at Cornell University as the salvation of the New York wine business. The Taylor Wine Company, the largest upstate winery at the time, told its growers to pull out all their native vines and replant with hybrids, sending the growers into economic shock.

When we first arrived back east, we assumed that we would settle in the Finger Lakes. After spending a couple of months exploring the region and finally learning some hard facts about grape growing, we concluded that there was something very unsettling going on there. Not only did it appear to be impossible to grow vinifera—especially red vinifera, which was what interested us—it was also as if the grape growers and winemakers of the Finger Lakes wine region were at war.

There were three men in particular—Dr. Nelson Shaulis, Walter Taylor, and Dr. Konstantin Frank—who were at one another's throats. They were leaders in the wine industry, but if they wished to be like Virgil or Columella, they fell far from the mark.

Nelson Shaulis (who actually looked a bit like the ancient Roman emperor Nero) was the professor from Cornell who showed up on our farm a couple of years later and cut himself slicing our buds with a razor. He was leading the charge to get vineyardists to plant hybrids, and didn't want to hear of anything else, unless it was the old Concord grapes that dominated juice production on the shores of Lake Erie.

Unlike Shaulis, Walter Taylor did have a personality of epic proportions. He was a grandson of the founder of the Taylor Wine Company and an employee of that company until he was fired for allegedly using company equipment and labor on his own vineyard, Bully Hill. Walter was unhappy that his family's company blended its wines with water and with bulk wine from California. To broadcast his disgust at the predominance of imported tank-car wine in Taylor's products, he displayed an empty tank car on his own property. There he started a winery with the motto "Wine without water" and planted hybrid grapes for the wine. At the entrance to his winery he mounted an oil painting that he had done of his grandmother, aged and nude. Walter was definitely into shock value. The Taylor Wine Company sued him for using his own name on the Bully Hill label, since the company owned its use in conjunction with wine, so he traveled around the country, wearing a mask like the Lone Ranger. His new label depicted a goat and the declaration "They Have My Name and My Heritage, but They Didn't Get My Goat."

Dr. Konstantin Frank, who also later visited us and called me a dirty farmer, was a brilliant and cranky old Russian of German parentage who had masterminded ways to grow cold-sensitive vinifera like Riesling when he ran the Crimean research station for the German occupation during World War II. Being more dedicated to grapes than to politics, he helped American forces restore Bavarian vineyards after the war. In the fifties he found his way to Cornell's Geneva Experimental Station, where he was given a job hoeing blueberries. Eventually, he got a job growing Chardonnay for the French Champagne maker Charles Fournier at Gold Seal. Gold Seal was the only winery in the Finger Lakes that had any success growing vinifera, partly because of its unique site on a sunny slope on Keuka Lake. Dr. Frank was so furious over being relegated to blueberry hoeing by Cornell that he mounted a campaign to convince the public that Shaulis's and Taylor's new hybrid plantings were toxic. They would cause birth defects, he claimed, saying repeatedly, "Look, now, you Americans are behind the moon."

The palpable rage of these men was enough to dampen our desire to plant much of anything anywhere near them, however personally cordial they were to us. Besides that, it took only a few tasting room visits to set us back on the vinifera trail. Maybe we could have found a site to plant vinifera in the Finger Lakes, but after traipsing for weeks through places with names like Frosty Hollow and Windy Ridge, we were convinced that the region was too cold for us to grow the varieties we liked. More than anything we wanted to grow Cabernet Sauvignon, which is a very late-ripening variety even in a warm region.

We spent nearly three months making forays into the lake country, putting pins into topographical maps and exploring tumbled-down dairy farms. We visited wineries in the Hudson River Valley, and Alex scouted out land in Virginia and Massachusetts, too. We were about to give up when John Tomkins, a professor of small fruits at Cornell in Ithaca, where (we thought) the grape wars were not personal yet, heard of our search and offered Alex advice: "Go east, young man."

Tomkins had been involved in a table grape project on the North Fork of eastern Long Island, helping a Cornell graduate named John Wickham. He had seen how well certain cold-sensitive grapes had done there and suggested we check it out. My parents still lived on Long Island, about sixty miles from Wickham's farm. We had planned to go to Long Island to visit them for Thanksgiving anyway, so we made an appointment to see Wickham on the day before the holiday.

Wickham greeted us cordially and drove us the short distance from his farm stand to Peconic Bay. He wanted to show us how that shallow body of water creates a long, warm ripening season and moderates the chill of the Atlantic Ocean, which lies on the other side of a parallel finger of land known as the South Fork, or the Hamptons. I listened to his words and looked at the marsh grass bordering the placid waters, which receded into a pastoral horizon punctuated by a white church steeple.

In my mind, I was back in about 1955, sitting behind my mother in Old Trusty, our Dodge sedan, with my father at the wheel and my brother and sister beside me. We were driving east on Route 25 as it curls down the hill from Laurel Hollow to the head of Cold Spring Harbor, with the steeple of Saint John's Church, where I was confirmed and married, to one side and the calm waters of the harbor on the other. As we did every time we made that approach, we waited for my father to start counting until, at the moment the harbor waters came into view, we all shouted, "God's country!"

The sight of the steeple also brought to my mind a Thanksgiving hymn that had always been a favorite of mine:

Come, ye thankful people, come;
Raise the song of harvest home.
All is safely gathered in
'ere the winter storms begin.

Its words filled me with a deep desire to find that safe harbor, that "harvest home." In the emotion of that moment I sensed my father's deep attachment to that part of the world as his own home. He had been sent away to boarding school when he was six, and from then had been in Cold Spring Harbor only on summer vacations and while on leave from fighting in World War II. Everything he loved was there—family and friends, sea and sky, and escape from the desperate realities of a world at war. The view I now saw with John Wickham was so similar to "God's country" that it resonated within me even though I had no consciousness of the effect it was having on me.

For the nearly three decades we owned the vineyard I would staunchly maintain that the reason Alex and I agreed to settle on Long Island had nothing to do with my having grown up there. In fact, I said, I wanted to be independent of my family. I insisted that the real reasons we were there had to do with hard scientific facts. Looking at statistics, we found that the North Fork compared favorably to the Finger Lakes, the Hudson Valley, Virginia, and the Connecticut shore. We could get a crop on the North Fork in three years instead of four. Here, in contrast to those other places, our vines wouldn't be killed to the ground in winter, so we would have a predictable crop every year. The pests that had devastated early attempts to grow vinifera could now be controlled with specialized pesticides and grafting. There was enough rainfall that we wouldn't need irrigation. There were none of the rocks and boulders so common in upstate territory, nor were there streams or heavy clay soils. Frost and drainage wouldn't be a problem, so we didn't need to find a slope or install drainage tiles. There would be no need to carry our soils back to the top of the mountain every year, since there was no mountain. The growing season was fully a month longer than in the Finger Lakes, making it more likely that we would be able to ripen the late red varieties. We were eighty-five miles from Manhattan, one of the world's biggest wine markets, and on a large island populated by prosperous people who vacationed on the east end. John Wickham told us that because there had been few farm-to-farm sales lately, we would probably not have to pay for any barns that might be on a site—they had no value to the typical developer. If worst came to worst and our vineyard didn't work out, we'd still own real estate that would not lose value, being located in a geographically limited resort area.

While it's true that Wickham was the person who showed us the benefits of the North Fork, he did not suggest we move there to plant grapes. On the contrary, he said, "Listen to me well. This is a wonderful place to farm, but I advise you not to do it. You have no experience, and it is far too risky. You have no idea how expensive it is to develop a new crop in a region. Don't be pioneers. Pioneers always pay twice. You'll end up with arrows in your backs."

It's not that we ignored John Wickham's words. They were sort of a dare. Wickham was a pioneer himself. He had introduced many crops—peaches, cherries, cranberries, even rice—to the North Fork. As he and Alex conversed on the merits of soils and the risks of hurricanes and nematodes, I could sense the two men bonding. They stood there, arms crossed and feet spread wide, both well over six feet tall. A generation apart, they were matched in intellect and intensity. Wickham was a model of a gentleman farmer whose love of family, history, and the land was evident in the way his blue eyes took in the landscape as he spoke and his straight back tilted toward the earth as he walked carefully over the furrows. Never mind what he said; it was clear that he would be a mentor for Alex if Alex decided to farm on the North Fork.

And then for me there was the smell of the salt sea. Before driving back to Rochester we went to look at Long Island Sound, just a couple of miles north of Peconic Bay. As we looked across the sound toward Connecticut, I inhaled the sea air deeply. It was the same air, the same sound in which my maternal great-grandmother had taken her children and grandchildren skinny-dipping on the Connecticut side of the water. Every summer, in a sheltered cove where we were the only swimmers, my family and I had done the same with my grandparents. The smell of that sea air brought back all those feelings of freedom without shame. Multiple generations with enough modesty not to care about nakedness had slipped into the clear, dark water just as the sun emerged in the summer dawn.

Alex agreed with me that there was no point in trying to grow grapes unless the conditions were right. And they were obviously right on the North Fork. But how did he respond to the salt sea air? If I had to do it again, I would have questioned his gut response to this place that was so unfamiliar to him. He had spent his summers water-skiing on the fresh waves of Canandaigua Lake upstate. The smell of the air and the light were different there. After we settled on Long Island, every once in a while, he would look at the Long Island landscape and say, "It's so flat."

We went back on my Christmas break, determined to find farmland for our venture. I began to appreciate the bare simplicity of the plain white cottages and farmhouses that lined up squarely along the highway. I liked the commonsense look of the shingled barns, with their peaked roofs and angular additions. However austere the Yankee architecture was, it played up the glory of the unobstructed sky and broad horizon. Sunlight shimmered off the water and reflected again off the leaves of grass in field after open field.

It didn't take long during our exploration of farm properties to discover that we were short of the money we would need to start our venture. I thought that since we wanted to do it, we could do it. It was that simple. Alex was more realistic; he started looking around for investors. Luckily for us, Alex's second cousin Bill Chapin got wind of our vineyard search. He had inherited some money from his father, who had died in a plane crash, and he was looking for an investment.

Bill didn't care much about grapes or wine; he just wanted to do something interesting with his money. That suited us fine. What could be better than a silent partner who lived in Rochester, five hundred miles away? Besides, Bill was an architect, and all of us thought that there might be the chance for him to design something exciting on the property once we made some money.

Bill and Alex worked out a partnership in which both would invest the same amount of money, but since Alex would be actively running the company he would be president and own fifty-one percent; Bill would be vice president with forty-nine percent. I didn't have any money to invest, but I was made secretary-treasurer since I would be working there, and we needed another officer in order to form a corporation. Alex turned his inheritance into cash and went back to find a farm that we could afford to buy. It would have pleased his grandfather to see his financial legacy turned into a new family farm.

After what seemed like an endless search, and a couple of bad experiences making bids on properties that the owners didn't really want to sell, we discovered a suitable farm in the village of Cutchogue, only a few farms away from the Wickham farm. We first saw it on a miserably wet January day, which we took as an auspicious sign. Don't they always say that buying property in bad weather is good luck? The farm was bordered by woods, one of the few parcels with buildings set off from the road, and it had a large potato storage barn that we could imagine converting into a winery.

We didn't spend much time examining the place in detail.

Since most of the farms we had looked at had no buildings at all, it seemed a miracle that there was a tiny farmhouse we could make our own. Could we move in and patch up this decrepit old place? Well, it sure beat the sod huts of the western pioneers. It was more spacious than the back of a Jeep, and it had marginally more daylight than the garage apartment we'd been renting in Rochester. Besides, we had run out of patience in our search. So we bought it.

5

Mother Nature

WE HAD BEEN AT OUR VINEYARD FOR LITTLE MORE THAN A year, and baby Anne was a few weeks old, when I started wearing her like a sweater. Centuries of farming mothers had worked with their babies strapped to their backs, papoose-style, so why shouldn't I? We couldn't afford a baby-sitter, anyway. Alex's mother gave me a patented corduroy carrier, a Snugli, which I treasured because it adjusted to the size of the baby and was soft and comfortable. I got so used to having Anne on my back that I'd forget she was there. One afternoon, when I went to get her up from a nap in her bedroom, she wasn't in her bassinet. I ran around the house in a panic, unable to think of where else I might have laid her down. Finally, when I passed a mirror, out of the corner of my eye I found her, still snoozing on my back.

She was on my back on a damp October day in 1974 when we picked our first crop. It's unusual to take a crop from two-year-old vines. Two-year-old vines are supposed to be making roots, not fruit. We knew that, and yet when we saw tiny clusters on our Pinot Noir and Cabernet Sauvignon vines in their second leaf, we didn't have the heart to drop them on the ground.

In May there had been half-inch-long embryonic clusters that became visible as the first leaves unfolded to expose them. These were really flower buds. At this stage, around the stem I could clearly see a double helix of buds that would become berries once they were pollinated. Surely God favored the grape over something as clumsy as an apple or a peach, which is ill supported by its tiny stem.

In June the buds burst open. Botanists call the grape's blossom "sexually perfect," as it contains both a stamen (the male, pollinating part) and a pistil (the female, receptive part) but no petals. It doesn't need petals to attract any bees because it can pollinate itself. It's just like a botanist to think that such an arrangement would be perfect, I thought. These flowers may not have looked like much, but in the evening when we went walking in the vineyard we smelled their ethereal fragrance.

After the pollen dropped from the stamen onto the pistil, causing the fruit to set, the berries quickly swelled. Nearly every vine had a fist-sized bunch of grapes on every shoot coming out of the trunk. They were only about a foot off the ground, so they'd be hard to pick, but we couldn't see any good reason not to let them ripen. Why not take the fruit and sell it? The vines were so vigorous, surely they could ripen this itty-bitty crop.

There would be too much fruit for us to legally make it into wine for ourselves. The federal government only allowed a "head of household"—meaning a man—the right to make two hundred gallons for private household use. Since we didn't have time to gear up with equipment for a commercial winery, we decided to sell the crop. Getting a little cash for our efforts sounded like a good idea.

Before we bought our farm we had met a graphic designer named Louis de Pew, who had started a small winery on the Hudson River. He had lived in France and had tried, largely unsuccessfully, to grow vinifera on the Hudson—the soils are too heavy and the winters are too cold—so we thought he might consider buying our fruit. We offered him both Pinot Noir and

Cabernet Sauvignon. (The Sauvignon Blanc was still recovering from having been attacked by rabbits the year before, so it didn't bear a crop.) Sure enough, Louis was interested in our fruit, as long as we trucked it to his winery.

As the summer of 1974 progressed, I watched the fruit swell, turn color, and ripen. September came, and the fruit got sweeter and sweeter. All the berries turned a deep blue-black, and we decided that it was time to pick.

Pinot Noir is the traditional grape of Burgundy, and Cabernet Sauvignon is the traditional grape of Bordeaux; since Burgundy is colder than Bordeaux, it stands to reason that Pinot would have to ripen sooner to escape frost in those more northern parts. We knew that these varieties would not be perfectly ripe at the same time, but when we discussed it with Louis, he had agreed that there would be so little of each—maybe under three tons—that we might as well plan to pick it all at once.

We scheduled a picking day, bought some bushel baskets designed for harvesting beans, and went out to pick. There was a cold fog that almost coated the earth, dulling the colors of the vines' leaves and fruit. I had a sense of anticipation: This is it! This is what it's all about! At the same time I felt awkward. Here was yet another task—like planting, hoeing, pruning, and tying—that we had never done before. These tasks aren't hard, but they still have to be done right. What do you do when there is no one to show you the best way to go about it? It was like childbirth, come to think of it. The job must be done, so you just get started and take what comes. You have no choice.

I wanted to memorialize the event—or maybe I was stalling—so while Alex set out baskets, I carefully, reverently snipped a cluster of Pinot Noir. Our first harvested fruit! It was perfect—smaller than a tennis ball, but crammed with so many berries in the cluster that they were fighting one another for space. I triumphantly placed this first cluster—Long Island's first vinifera harvest—in a clean jarful of formaldehyde. It was my trophy.

Maybe it was weird to keep this pickled cluster. Back in my high school years I had been fascinated by the jars of specimens in formaldehyde that lined the walls of the biology lab and gave the room its pungent aroma. I thought they were tokens of a quest for knowledge, and I loved the time I spent in class doing dissections. What I hadn't liked was the way my biology teacher had played favorites by giving one girl a special foot-long Australian worm to dissect, while the rest of us got ordinary earthworms. Keeping this cluster was a way to tell myself that my interest in these grapes, scientific or not, was independent of any reward—it was its own reward.

Back in the vineyard, Alex joined me in picking. He was on his hands and knees, almost lying down, to get his tall frame low enough so that he could see the crop. Meg was there, too, as well as a young woman we had hired for the day. She was an attractive, educated woman who had been living in her car with her young child since her husband had abandoned them. I learned that she had been leading an affluent life until her son was born. Her husband couldn't stand the competition and took off. As a single mother on the North Fork, she hadn't been able to find work that would allow her to care for her son. I was stunned to learn this. I felt as if I had a rock in the bottom of my stomach. If it happened to her, it could happen to anyone. But not me, I thought. That would never happen to me.

Her son came with her, of course. I wondered what other pioneer mothers did with their children while they worked in the field. Did they succeed in teaching them to be "seen but not heard"? An infant can be carried, wailing or sleeping. An infant can be suckled in the field. But an older child was a constant interruption. I would have to figure this out for myself before too long.

We cheered when the last grapes were picked and stowed in the back of Mike Kaloski's pickup truck, which he loaned us for the occasion. There was no time for a party because we still had to make the long drive up the Hudson River to the winery. It was almost dark when we got there. Our excitement was intense as we pulled into the hilltop winery. Louis de Pew's grown son, Sam, greeted us with a smile. We were all congratulatory until Louis came out of his house and said, "I spoke to my friends in France this morning, and they said I must not use the fruit of two-year-old vines. The French would never do it. I have to refuse the shipment."

I stood there, embracing my squirming baby. My back was screaming in pain from picking and loading our precious fruit. I looked at Alex, who seemed livid. What were we going to do with this fruit now that it was picked? Restraining my impulse to cry in fury and disbelief, I walked away from the men and looked at the fruit that was waiting by the winery's crusher, fruit from its own vines. It looked terrible. Some had mold on it, and clearly half the berries had been shredded by birds. I nibbled on a few berries. They tasted sour and flavorless to me. Alex came and inspected it, too. Was Louis really going to decline our fruit when his own looked like this?

Baby Anne and I kept out of the way as I watched the three men, bodies rigid and hands gesticulating, discussing the subject. Finally, Alex came over and told me that they had struck a deal. Louis would accept the fruit, but for much less money than we had agreed on. It wouldn't cover our expenses, but at least the fruit wouldn't be thrown away.

Embarrassed and upset by his father's behavior toward us, Sam invited us into his loft apartment and cooked us dinner to try to make up for the fiasco. A short time after this harvest, he left his father's winery and started his own, a successful one in another wine region. The next year, Louis released the wine. I don't think I ever tasted it. It was wine made from sour grapes.

By the time we brought in that first harvest, I had become callused and muscular. All summer I went barefoot and I was lucky that I didn't cut my feet on the shards of old pottery and bits of rusted farm equipment that kept emerging from the depths of the soil as if they were another crop. It was interesting to pick up these fragments from the lives of other families who had worked our farm. Examining a bit of a plate that was decorated with pink flowers, I thought about whether they had bought it after a good harvest, or if it had been part of a farmwife's dowry. After a day's work, who had smoked the clay pipe that I found in the field? Had he sat by the stove and talked with his wife about how to pay for seeds? What about the bit of horse harness that came up out of the ground—was the horse whipped if it was too slow?

As I hoed a row in my halter top one hot day, thinking about all the people who had hoed the same piece of land before me, I noticed a Cadillac idling in our driveway for a good fifteen minutes. When I walked over to it to see if the driver was lost, the door opened and I recognized the former potato farmer who was now our insurance agent. He had grown up farming not far from here. He stood on the packed part of the road—to keep his white bucks clean, I supposed—and greeted me. "You better be careful not to work so much," he said, with a leering glance at my arms. "A girl shouldn't have muscles like those."

I liked my muscles, and I liked Alex's strong and slender arms, too. For the first few years, he worked with me in the field. He did the tractor work, too, until one day when his back went into spasms from the impact of the tractor's vibrations. After a couple of nights being treated with muscle relaxants in the local hospital, he enlisted Charlie to take over most of the tractor work while he retreated to his study to figure out how to proceed with the vineyard. Charlie became a quiet but steady partner, coming and going from time to time, but there when we needed him, like an anchor to windward, over the next twelve years. Meg went off to college and then got married. She had been with us only sporadically in the first two years of the vineyard, but she had embraced the work, literally. She couldn't walk through the vineyard without getting covered in dirt, or bake bread without getting covered with flour. When she left, I missed her company—her humor, her honesty, and the way she put a new twist on everything she did.

In the vineyard we were still relying on Virgil, Columella, and *General Viticulture* to teach us what to do. The California text showed training systems in every theme and variation—the Guyot, the Double Kniffen, the Umbrella, and so on—in two dimensions and black and white. The tangled mass of growth in front of us was like a jigsaw puzzle, and I was never good at jigsaw puzzles. But then I compared the job to cutting hair. I did cut Alex's hair, not because I knew how, but because we couldn't afford the barber. "Are you sure you want me to cut your hair?" I would ask Alex before I started. He always said yes. His hair grew quickly, and who was going to see it, anyway? Well, then, I thought, I might as well start cutting the vines in about the same way, trying to even things out to end up with some kind of reasonable shape.

For every pound of pruning wood, I read, we were supposed to leave a certain number of buds on wood that had grown the year before. You can't know how much the wood weighs until you cut it, and, like hair, once it is cut, it can't be put back. For weeks, we counted all the buds and weighed all the wood. Dave Mudd came along and made the chore more sociable. None of us knew what we were doing.

It was a relief when Dr. Shaulis, the professor from upstate who had cut his finger testing our vines for winter kill, offered to come back with a crew to make a formal pruning and training trial. He and John Tomkins, the Cornell professor who had suggested we come to Long Island, had secured funds from the state to bring four or five experienced vineyardists from the research station in Geneva, New York, down to our farm. They would come at least four times a year to prune and train several carefully selected rows of Cabernet Sauvignon and Pinot Noir. There would be three levels of pruning: severe, moderate, and light. They also set up a trial of Shaulis's pet invention, an elaborate trellising system called the Geneva Double Curtain (GDC). At harvest, they would weigh all the fruit from each plant, count the berries, and test the sugar, acid, and pH of the fruit from each trial. The idea was to discover what treatment would yield the ripest fruit.

The pruning crews brought a festival atmosphere to the vineyard. We felt connected to the wine world as we listened to their stories about grape growing in their part of the state. When harvest came we took the extra time to help them with all the complicated weighing and counting. After five years of the experiment, the results showed that at every level of pruning, the fruit ripened well. There didn't seem to be an advantage to pruning severely, and in fact the most severely pruned plants grew so densely that their leaves shaded their own fruit clusters.

These results were exciting because now, in a state-sponsored scientific experiment, we had proven that, contrary to three hundred years of failure on the East Coast, *Vitis vinifera* could be grown and ripened on Long Island. We eagerly awaited the publication of the results of this experiment, but it never happened. We were told that Dr. Shaulis was retiring and his replacement had reservations about the data. With a shrug and a sigh, we let it pass. A few years later, when other growers came to the island to plant vinifera, we were flabbergasted to see that, instead of heading their vines at three or four feet tall—the way we and most growers of vinifera in other cool climates all over the world trained them—they used Cornell's recommendations for native grapes in training their vines, heading at six feet. It seemed obvious from the results of our experiment that the high cordon was entirely unsuitable for upward-growing Cabernet and Chardonnay. We wondered why they ignored everything that our experiment had demonstrated.

Before I had any thoughts of righteous indignation about grapevine training—before the pruning experiments even began—there were many days and nights when my attention and all my heart were focused on our new little family. The winter that Anne was born she slept in her bassinet next to my side of the bed. When she woke up during the night I would scoop her up and nurse her while I lay on my side. Sometimes, Alex would spoon around us. It was a time of bliss.

Lying like this, I would listen to the wind and its varied pitches against the trees and the roof. Often it was silent. One night, instead of hearing a slight clang as the wire from the old TV antenna rhythmically hit its support, I heard a gong, gong, gong. The wind had risen ferociously, turning the cable into an alarm. Outside, the gentle snow that had been falling had intensified into a blizzard. I thought of Laura Ingalls Wilder in *The Long Winter,* when her family was snowed in on the prairie for months. Laura described how one night she awoke in darkness, feeling strangely warm. Snow had come through the roof, completely covering her over her buffalo robe. At a perpetual thirty-two degrees, the snow felt warmer than the air. This image of an unexpected peace comforted me as I listened to the storm. They had survived, and so would we. I cuddled with my little family and fell asleep again.

The blizzard continued all the next day. We really were snowed in. Although we couldn't get out of our driveway, we still had heat. It wasn't exactly like being on the prairie. Nevertheless, on the third day of being snowed in, I was beginning to worry about our food and diaper supply when I heard a new noise—heavy machinery. There was John Wickham, riding his Caterpillar bulldozer all the way from his farm, across the back fields, to see if we needed anything. He moved enough snow around so that Alex could get our own tractor out of the barn, and the two of them went riding back to town with a shopping list.

Through that and every other blizzard that hit the North Fork, I never felt fear. Of course, there was excitement, and a certain trepidation as the cold wind penetrated our little house, but major snowstorms were rare, and they were always followed by several days of brilliantly sunny skies. Those were the hot-chocolate days. All winter we worked outside, pruning the vines. For a time I had Anne on my back, her cheeks pink and her hands and feet protected by her father's woolen socks. Later, Anne went to a baby-sitter for part of the day, so that I could keep working. Eventually, after we had planted more acres of vines, we hired a couple of people to help us prune and tend them. One of the pruners crocheted each of us little nose warmers made of reddish-brown wool, so that we looked like Irish setters, especially Alex, who wore a hat with fur earflaps. In a winter that brought three-foot drifts of snow into the vineyard, the crew and I had so much difficulty walking from vine to vine that we dropped to the ground and rolled, hooting and howling, from one plant to the next. On Thursday nights, the crew would come to our house to watch *Kung Fu* on our twelve-inch TV and drink a jug of CK Mondavi red Zinfandel. *Kung Fu* starred David Carradine as a kung fu expert, who as a young boy is called "Grasshopper" by his mentor, a monk. While we watched it purely for entertainment, I was impressed by the strength that Grasshopper took from his spirit.

The sultry summer days of August always brought lightning and thunder. With its miles of galvanized wire trellises, the vineyard was a dangerous place to be. After seeing an entire row of vines fried by a bolt that struck a wire and burned everything that touched it, we learned to run from the area with the first flash.

Our insurance guy (the one with the white bucks) told us we could save money if we installed lightning rods on our house and barn. We got an estimate for the rods but decided we couldn't afford them. That night, a zigzag line of electricity hit the roof followed by a crack of thunder. We could smell the roof tiles burning. God had spoken.

Fortunately, it was just a word, not a sermon. The fire didn't spread, but we did buy lightning rods for the buildings. Even then, electric storms were frightening. A rod was hit once when I was standing in the kitchen, talking to Alex. The powerful electric charge traveled down the thick wires that grounded it. I was so close to it, on the other side of the wall, that I leaped into the air and found myself in Alex's arms. His hair was literally standing up, and I could feel my own scalp on alert.

I learned to deal with the wailing of the baby and my own fear during these storms by banging on pots and pans. Little Anne and I tried to make more noise than the thunder, clanging pots in time with the beating of our hearts, until the force of the storm passed.

We would often lose power when we had a storm. Our part of the world had had electricity for a long time, and the wires and transformers were very old. Trees had grown densely around power lines, making them vulnerable to breaking branches. Still, I felt safe in the average thunderstorm. Hurricanes were another matter.

In 1972, when we first met John Wickham, we had talked about the potential for a severe hurricane on Long Island. He told us about the infamous hurricane of 1938, when an unexpected blast had roared up the East Coast with hundred-mile-an-hour winds, slamming into the Hamptons and decimating the waterfront. Then, he mollified us with the statistic that hurricanes came on average once every sixteen years. We could live with those odds.

I remembered that just before my seventh birthday in 1954, Hurricane Carol, the most dangerous hurricane since 1938, had landed ashore at high tide near where I lived in Cold Spring Harbor. I distinctly recall standing in the front yard with Wendy, watching the huge trees moving from the force of the wind, not just in their branches but in their trunks, too. As the wind picked up our mother made us come inside, where I stood at the picture window to watch. The leaves had begun to color, and I was impressed by how quickly they began to litter the yard. I loved jumping in piles of leaves, and I thought about how much fun I'd have with them when the storm ended. The next day, my mother drove us to the harbor to see how high the water had come. It still covered everything at the foot of Snake Hill, two days afterward. There were large boats that had been tossed up onto the land, and various docks were littered along the beachfront.

School started just after that, and the next time I saw the beach was the following summer, when everything had been repaired and looked perfectly normal. After that I just didn't feel the sense of alarm I knew I should have had about hurricanes. From what I had seen, it hadn't taken much to repair the damage.

In early August 1976 we heard warnings for Hurricane Belle, which was barreling up the coast early in the hurricane season. Our twenty-seven acres of vines were young but they were in full leaf, with an important crop just starting to turn color. This was truly frightening. There was no way to anticipate what the storm might do, and nothing to do about it but hunker down.

Belle turned out to be devastating for the peach and apple growers of New England, but it passed over us so fast, and our vines were so flexible, that we thought we weren't damaged at all. Grape clusters at this stage, more than a month before harvest, are as firm as the polyurethane of a dashboard, and their stems are as tough as nylon. I wondered what it would have been like if the storm had hit us at harvest, when a firm shaking would have scattered the berries on the vineyard floor.

Two days after Belle went on to batter Cape Cod we saw the youngest vines' leaves turn brown and shrivel up. The storm's gusts had carried salt air inland, scorching any tender vegetation. The older vines had enough leaves that it didn't matter; the interior of the vineyard canopy wasn't hurt. All the vines planted in 1974, however, were defoliated. We had expected to take a crop from them, and while the fruit remained on the vines, there weren't enough leaves left to ripen it. Our financial plan went out the window.

We often experienced tropical storms, which could be just as frightening as hurricanes. There was one nor'easter that kept up for several days. While our farm appeared to be level from a distance, there was actually a gentle slope to the west where the red grapes were planted. As this storm persisted, the soil from the top of the Cabernet vineyard began to wash away in the night. In a torn slicker, Alex went out and tried to redirect the eroding soils with sandbags, brush, and anything else he could haul into the path of the flood. I tried to help him, but I wasn't big enough to wrestle the bags into place. When the next day dawned with fresh, clear air, and the vineyard sparkled despite the deeply gouged trenches of outwashed soil that reminded me of a map of the Mississippi Delta, I felt that the desperate efforts of the night before had been ridiculous and futile. But even though the winner of a battle between man and nature is a foregone conclusion, we engaged in it again and again.

Every year from then on, we watched hurricanes as they formed in the Caribbean and made their way up the coast. Usually they went inland to Florida or Cape Hatteras, or blew out to sea. In a perverse way, as much as I wished the storms would dissipate, at the same time I wanted them to hit us. It was a kind of dare to the heavens to show me their worst, so that I could prove myself invulnerable. I wanted to feel the extremes of nature—to be blown over by the wind, to be covered in a snow bank, to faint with the intensity of the summer sun. Then, I could run to the shelter of my home, which would feel all the safer in the brunt of the storm.

Many years passed before the serious consequences of a hurricane changed my attitude toward the storms. In late September 1985, just after we had framed an addition to our farmhouse with a new fireplace standing unprotected at the end of the structure, Hurricane Gloria smashed Long Island with eighty-five-mile-an-hour winds. We had already picked our Pinot Noir and started on our Chardonnay when we started watching reports of the storm coming up the coast. The day before it hit, we decided to stop picking Chardonnay and switch to Merlot, which was also ripe. Because of the size of the crop on the vine, and the way it hung down from the trellis, the Chardonnay was easier to pick, and so we could have gathered more of it in that day's harvest. The two kinds of wine could be sold for about the same price, but Chardonnay would be ready to market several months sooner since it needed less time in barrel. I cared more about the Merlot, however. I felt that our reputation would be made with red wine, and Merlot was our best.

With the winds increasing to gale force, before the rain came driving down at a flesh-piercing angle, my crew and I speeded up our work in a lather of mutual cheerleading. Before all the crop was in, we had to retreat to safety, running alongside the tractor and throwing the picking boxes on the cart until most of the Merlot was safely stacked by the crusher.

Like Belle, Gloria blew through quickly, but the fruit was much more vulnerable. Fortunately, there was more rain with Gloria, so the salt damage was negligible. When the skies cleared, while the Chardonnay and Sauvignon Blanc grapes that remained in the field were lightly bruised on the surface, the angle of the storm had aligned itself with the trellis wires, so that damage was minimal. Still, the bruising could turn to rot if we didn't complete the harvest in a hurry.

Rejoicing at the sight of our largely unscathed vineyard, we were ready to get back to our harvest the day after the storm, but we could do nothing in the vineyard or winery until power was restored. All of middle and eastern Long Island had lost power. Without electricity, we couldn't run the crusher, the press, or the pumps. The crop sat in the field day after day while the power company focused its efforts on the communities that had larger populations than the North Fork. By the fifth day, we were panicked that our crop would rot in the field. It was time to press the Pinot Noir, too, and every day we waited would make the wine taste harsher.

I was fed up that our power company had neglected to trim trees and replace worn-out transformers for many years. I said to myself, "It's time to get on your pony and ride." I got in the car and started cruising the highway, looking for electrical utility trucks. Other power companies from adjacent states had been called in to help, and when I spotted a caravan of repair vehicles, I followed them to their substation, which was hidden in the woods. As I walked past the NO TRESPASSING sign and into the control building, I felt like a sheriff opening the door to the barroom where the bandits are hiding. With a woeful look on my face I explained to the beer-bellied repair team that if we didn't get power immediately, our harvest would be ruined and the power company would be responsible. "We aren't allowed to hook up any transformers that serve a population of less than two thousand," the chief said. I kept on pleading my case to this stone wall of a man. He kept shaking his head. I left in despair. Three hours later our lights suddenly came back on. The crew of a Pennsylvania utility company had come to our rescue because they wanted to see what a winery looked like. Some of my closest neighbors must have wondered why they had power, along with us, when the rest of the area remained dark for as long as six more days.

We got back to picking and the crop was saved, but what we hadn't reckoned on was the impact of the hurricane on our business. In any year, we didn't sell enough wine to make a profit until September, and October was our biggest sales month. That's when the greatest number of visitors came out to buy at the full retail price. But all of Long Island was repairing roofs and cleaning yards that October, and the business didn't materialize. We couldn't afford to finish building our addition; in fact, it wouldn't be finished for over ten years.

By November we didn't know if our business could survive. We decided to make an early release of our Pinot Noir, which was made in a light Beaujolais style to begin with. It would come out at the same time as the French Beaujolais.

We had to come up with a special name for it, and I thought it would be romantic to call it "Candlelight Burgundy." Pinot Noir is a Burgundy grape, and the wine was at least partially fermented by candlelight. Alex wanted to put "Hurricane Gloria" on the label, which I thought was a terrible idea. Who would want to buy a wine that was made from fruit battered by a storm? And why would we mention the storm if the fruit had been picked, as it had been, before it?

Alex and I compromised and used both names. On the day we

released our "Hurricane Gloria Candlelight Burgundy," *Newsday,* Long Island's all-powerful newspaper, wrote about it on page two. We sold over a thousand cases of it in a week, the most wine we ever sold in anything close to that amount of time. One man who had made a fortune repairing roofs damaged by the hurricane bought fifty cases for his customers, and judging by the numbers every woman on Long Island named Gloria—or her husband—must have bought a bottle.

When I was a child, outside my family's house in Cold Spring Harbor there was a hydrangea bush that had huge balls of turquoise blossoms. I loved these flowers, but my mother thought they were garish. I wondered how she could be critical of a beautiful flower, when she wasn't critical of the way I cultivated various insects in and around the house. She laughed with me when I showed her the daddy longlegs I found in the bathtub. When we saw a ladybug, we would make a wish together and chant the nursery rhyme about ladybugs, and I would fervently hope that the ladybug would indeed fly away home fast enough to save her children. Why did she leave them in the first place, I wondered? And where was mommy longlegs?

In the early fifties my parents built a small brick terrace where we liked to eat in the summer. The crumbs from our outdoor picnics brought ants to the terrace, and my mother let me try to train them to march single file toward some sugar cubes that I put out there. After a few days of organizing my ant colony I got tired of squatting on the terrace to watch them and went on to other games. When iridescent Japanese beetles invaded my mother's flowers one year she and I argued again about whether something so flashy could be considered beautiful.

I took my benign appreciation of insects to the vineyard. One of the advantages to starting a new crop in a region is that it takes time for the pests that like the crop to find it. Nonetheless, there were bugs everywhere. When the first leaves appeared on the vines, I noticed tiny spiders that wove filaments from tip to tip. As I worked in the field, I'd get covered by these webs, and by spiders, too, but they were too small to be a nuisance. The Japanese beetles that I had formerly admired were another matter, as they would use their pincers to latch onto my hair and often fell inside my shirt as I worked the vineyard rows. They did eat grape leaves, but most of the time there weren't enough of them to cause significant damage. And I still thought they were beautiful.

The legions of orange-and-black-striped beetles that both crawled and flew seemed to be more threatening. Before Alex learned that they were called Colorado potato beetles, he held a few of them in his hand, looked them in their multiple eyes, and told them that if they didn't hurt him, he wouldn't hurt them. The bugs apparently listened to him. After the Kaloskis' potatoes had been harvested, I saw their barn covered with beetles looking for food. They stayed out of the vineyard the way a cat stays out of water.

The potato farmers had created a problem by specializing in one type of crop. No matter how toxic a pesticide they used, the populations of potato beetles were so large that there would always be survivors that mutated and were resistant to the latest poison. We had potato farms on three sides of us, and they were sprayed regularly by helicopter, usually at six in the morning before the wind came up. On a lazy Sunday, we'd be lying in bed when the chopper overhead made our farm sound like a Vietnam battlefield, and its trail of poison left the air reeking like rotten garlic for the rest of the day.

My mother used to spray DDT in the kitchen, just to make things tidy. I loved the sweet smell of it and I would follow her around sniffing the cloud that came out of the shiny blue spray bomb. On our farm, I wasn't exactly thrilled with the idea of using pesticides, but I figured that it was a part of modern farm-ing. We had only been on the farm for a few weeks when the local pesticide salesman introduced himself and gave us both purple hats with the Benlate fungicide logo on them. My purple Benlate hat was my badge of authenticity, and I wore it like a cowboy wears a Stetson—all the time.

Although I wore the hat, I never did any spraying myself. At first Alex did it, and later his brother and others took on the job. The main thing we had to spray for was fungus, microscopic or-ganisms that cover the foliage of lilacs and zinnias, make black spots on roses, and cause closets to smell musty in the sum-mer. Fungus, especially black rot, downy mildew, and powdery mildew, is a threat in vineyards the world over. Even organic vineyards have to spray sulfur or other approved substances to control them.

Before we started the vineyard, we had seen the kind of dam-age a fungus can do. Upstate, we visited a vineyard that had been attacked by downy mildew. All the leaves and fruit were covered with a gray fuzz. The air smelled of vinegar and mildew. This same kind of mildew was one of the reasons that grape growing on the East Coast had been unsuccessful in the eigh-teenth and nineteenth centuries. It became a scourge in mid-nineteenth-century French vineyards, shortly before the invasion of phylloxera. Pierre-Marie-Alexis Millardet, a French farmer who owned a vineyard alongside a well-traveled road, wanted to discourage wayfarers from eating his crop. He invented a paste of blue-green stuff made of copper and sulfur that he painted on his fruit. Not only did it repel the picnickers; it also rid the vines of fungus. Thus was born the famous "Bordeaux Mixture," a fungicide that is still in use today. Millardet neglected to patent his paste, and when another grower claimed to have in-vented it, there ensued the sort of battle the French seem espe-cially to enjoy.

Besides the Bordeaux Mixture, which can burn the vines'

leaves, we also had at our disposal Benlate, Ferbam, Captan, sulfur, and later other formulations. Alex studied and got a license to apply pesticides to our farm, learning to his relief that these fungicides, unlike the pesticides our neighbors were using, are less toxic than aspirin. He always followed the guidelines published on the labels of these things, as well as the guidelines created by Cornell University for New York's grape industry. We had to spray much less frequently than vineyards in many parts of the world. When we were able, we bought a bigger tractor that had a cab with a spray-safe filter, air-conditioning, and a radio, so that spraying could be done in relative comfort and safety.

For me, working out in the field for the better part of every day, the residue of these chemicals was something I chose to discount as an unavoidable nuisance. Some days, it was impossible to pull weeds, position shoots, or thin the crop without coming home smelling of sulfurous black Ferbam. It covered my arms, but I got pretty good at working with them extended so that my body stayed relatively clean. I didn't wear gloves. They made me clumsy and got in the way of my ability to feel my work. I could wash off the Ferbam, and by harvest time whatever spray had been on the vines had washed away as well. But I couldn't wash off the juice stains from the fruit, which made my hands purple and then gray-green as soon as they came into contact with soap.

Fungus was a bad problem, but birds were worse. At first, in early spring when the robins returned to build their nests, I enjoyed all the birds that thronged the vineyard. I'd awaken to the melancholy cooing of mourning doves and engage the mockingbird that sat on our roof in conversation, repeating whatever he said and trying to get him to copy me. There were owls at night, and chickadees, sparrows, and blackbirds during the day. There was a bird that sounded like a flushing toilet—Meg and I called it the toilet bird. The woodpeckers drilled holes in our roof, and there were crows that arrived in a black cloud, all at once. The crows reminded me of when I was a child and my father would throw open the upstairs bedroom window if they landed on our roof. He would fire a shotgun to disperse them and say somewhat gleefully, "Caw, caw, says the crow; spring will come again, I know."

My attitude toward birds changed radically in our third season, when birds began to attack our fruit as it reached ripeness. After bloom in June, when the berries are pollinated, they increase in size but not sweetness until one magical moment around early August, when they reach *veraison*. The clusters turn color, from green to some shade of purple, red, black, yellow-green, or gold. Suddenly, within days, the sugar increases to about twelve percent, with another point gain about every third day after that. Birds that eat grapes will nibble at fruit with less than seventeen percent sugar, but once the sugar gets higher than that, they become voracious. Their screeches of glee alert other birds, and pretty soon it's like the Hitchcock movie.

The first grapes to hit the seventeen-percent mark were the Pinot Noir. Pretty little yellow finches would hover over the tight clusters, jabbing their pointy little beaks into the softening pellicles. Juice would run all over the cluster, attracting yellow jackets and mold. I wasn't so crude as to say it out loud, but I agreed with Alex that they were "fucking finches."

Following the finches came the *Turdus migratorius,* the robin redbreast, the aptly named plucker. Phalanxes of robins came zooming into the vineyard from their perches in the woods and the hedgerow, fitting whole berries into their craws. The starlings that came next just added insult to injury. Our vineyard was on the Atlantic flyway, which meant that every migrating bird on the East Coast flew over us. Remember the nursery rhyme about "four and twenty blackbirds baked in a pie"? It sounded like a good idea to me. And don't think it hasn't been done. Bordeaux has no more songbirds. But in the United States all of these birds except the starlings are federally protected species.

The red Toyota that had replaced our worn-out Jeep became the laughingstock of the birds as we chased them through the vineyard, honking up and down the rows from sunup to sundown. The birds would fly up for a few moments, then settle back down again. Over the years, we tried adding various sounds to our bird-chasing routine. We had a radio mounted to a pole, blasting rock and roll. I don't know how the birds felt about it, but to this day if I hear "Hotel California" or "If You Leave Me Now," I become a dangerous woman. For some reason I never got tired of "American Pie," and I don't think it scared the birds, either.

We got more sophisticated and added propane gas-fired cannons, flashing airport strobe lights, and dragon kites that were suspended from upright thirty-foot irrigation pipes. Finally, we added recorded avian distress calls like the ones we had first heard in the vineyards east of Monterey. They worked for a few days until the birds realized that nothing was actually being tortured. I anxiously wondered how that recording had been made.

Next, we bought eleven miles' worth of bird netting, which we draped over the rows individually. It was a huge task getting it installed. The birds would get trapped inside the nets, and someone would have to go out twice a day to push the trapped birds to the end of each row with a tennis racket, so that they could be released. It was an awful chore.

For the next few years, we had the luxury of using a new spray material that was already in use on other crops like corn and cherries. It wasn't toxic to birds, but they hated the taste of it. One nibble and they would warn their friends to stay out. It worked well, but it washed off in the rain and cost a fortune. After a couple of years its approval for grapes was revoked, even though it was still being used on other crops. We had already sold all our nets and couldn't afford to buy them again, so we had to go back to chasing the little devils.

For a short time, we had a theory that if we had enough cats, the birds would stay out of the vineyard. Iota was a pretty good hunter, and Eeper was, too. Cats were easy to come by. We adopted a black-and-white kitten whom we named the General for the way he marched around in a very un-catlike way. The name stuck even after we found out that he was a she.

Someone then talked us into adopting an orange cat that we named Aunt Bobbie after Alex's redheaded aunt. Too bad the cat couldn't make angel food cakes with chocolate whipped-cream icing like the real Aunt Bobbie.

While we were adopting these other cats, Iota gave birth to a litter of four: Fuzz, Buzz, Wuzz, and Wasn't. We gave away all but Buzzy, who was lazy and sadly ineffective as a hunter. In truth, none of these cats, nor their offspring, wanted to do any serious hunting. A mouse here, a vole there, maybe, but they were certainly unwilling to patrol the vineyard when there was a perfectly good stoop to lie under when it got hot, and a little girl to play with if things got dull.

After we got up to twenty-two cats, including kittens, we realized we had to cut back. We had all of the adults spayed or neutered and gave away the little ones. Still, people kept begging us to adopt more misbegotten cats. "You have a farm," they would argue, "there's plenty of room for Fluffy." A few times, we caved in, and so it happened that we had a cat who gave birth in the house and then moved the kittens to the favorite cat spot, under the stoop. We knew there were four kittens, but we couldn't get at them. When the kittens were just a few days old, we were awakened by horrifying shrieks in the middle of the night. I ran to the stoop to see the frantic mother cat darting between the yard and the stoop. For four nights, the shrieks were repeated. The next morning, the mother cat came into the house, her teats swollen with unconsumed milk. The kittens were gone. A mother raccoon had built a nest in the old chimney of the potato barn, and every night she had stolen a kitten to feed her own babies. After that, I lost my desire to have kittens around. Making wine and taking care of my family were much more fulfilling.

Springtime at the Hargrave farmhouse in Cutchogue. *Courtesy of Hargrave Vineyard Archives*

Louisa and Alex proudly show their young vines. This photograph appeared in the *New York Times*. Summer, 1973. *Photograph by Barbara Delatiner*

Louisa and Alex inspect a shipment of live grapevines in front of their farmhouse after the vines were threatened by frost. Spring, 1974. © *Newsday, Inc.; reprinted with permission*

Louisa and Alex with newborn daughter Anne. August 1974.

Photograph by Mary Lindsay; courtesy of Hargrave Vineyard Archives

Moving a stainless-steel tank into the winery. Summer, 1975. *Courtesy of Hargrave Vineyard Archives*

One more visit to the vines before Anne's bedtime, and the cats are delighted. Summer, 1975. *Courtesy of Hargrave Vineyard Archives*

Anne tests the fruit during the 1975 harvest.
Courtesy of Hargrave Vineyard Archives

The Hargrave family
in the wine cellar.
Fall, 1977.
*Courtesy of
Bruce Ando/Ando Photography*

Vineyard worker Steven Graves brings the Cabernet Sauvignon to the winery. Harvest, 1977.

Courtesy of Bruce Ando/Ando Photography

Alex's girls, *from left:* his grandmother Catherine Hargrave; his mother, Betty Hargrave; Anne; Louisa; his grandmother Florence Crouch. Summer, 1976.

Courtesy of Hargrave Vineyard Archives

Louisa, Alex, and crew, all bundled up and pruning Cabernet Sauvignon vines. Winter, 1978. *Courtesy of Hargrave Vineyard Archives*

Alex and Zander work together by the winery. Fall, 1979.

Photograph by Jim Mooney; courtesy of New York Daily News

Louisa cooking at the Garland stove while she and Alex discuss the events of the day. Fall, 1979. © *1979 Newsday Inc.; reprinted with permission.*

Anne with the sheep, Phyllis and Lucy; the horse, Silhouette; and the dog, Zeus, hidden in the grass. Summer, 1984. *Courtesy of Hargrave Vineyard Archives*

Zander inside a fermenter, after shoveling out tons of Merlot pomace. Harvest, 1997.
Courtesy of Hargrave Vineyard Archives

Louisa fishing in New Zealand. February 2000.
Courtesy of Lisa Van de Water

6

Fruition

WELL BEFORE WE EXPERIENCED ALL OF NATURE'S TRIALS, AND after the fiasco of selling fruit to the winery on the Hudson, we needed to prepare to make wine ourselves. Nineteen seventy-five would bring our first commercial vintage. We were excited to be doing it all ourselves, but it would have been nice to have a little more assistance. What sticks in my mind about the harvest that year was the way that Alex repeatedly muttered the tale of the Little Red Hen. The Little Red Hen wanted to make bread, but no one would help her. "Who will help me pick the corn?" she asked. "Not I," said the cow. "Not I," said the sheep. The other animals all made excuses, too, so she did it herself. "Who will help me grind the corn?" she asked. More excuses, so she did it herself. This went on through grinding, kneading, and baking the bread. Finally, she asked, "Who will help me eat the bread?" and everyone said, "I will!"

The agencies that were created after Prohibition to regulate wineries were not designed to help us, and they often worked at cross-purposes. The federal government and various state legislatures saw alcoholic beverages as a source for tax revenue, rather than as a benefit to the economy.

As applicants for a winery license from the federal Bureau of

Alcohol, Tobacco and Firearms (BATF), both Alex and I had to be fingerprinted like common criminals. Down at the local police station a cop snickered as he took each of my fingers and rolled it over the inkpad. With a stern grip, as if I might run away, he then directed my paws over to a form where he rolled each finger the other way. When John Wickham pulled Alex aside at the post office to say that some federal agents from the BATF had been out asking questions about our sex life, we understood why some of our neighbors had started looking at us sideways. His response—"How should I know?"—had caused them to be even more persistent. They wanted to know if we were perverts or Commies.

We weren't perverts or Commies, but we were small-winery operators, which was almost as great a crime. Before the New York Farm Wineries Act was passed in 1976, every winery license applicant, no matter how small, paid the same fee and was subjected to the same level of investigation. Even after the act was passed, applicants were subjected to thorough investigation by the federal government. This was America! Why would anyone want to be small? It didn't seem to matter that the federal government was spending more money investigating us than it ever received in alcoholic beverage taxes on our wine.

While I was out tying vines and battling weeds that summer, Alex was consumed by the application business. The BATF, the State Liquor Authority (SLA), the Department of Agriculture and Markets, and the Board of Health all set hurdles at various heights. The worst part of it was, it was never clear what steps were really required. The governmental agencies didn't speak to one another. And they certainly didn't know what to do with a winery like ours, making under ten thousand gallons. Even with all the paperwork done, the SLA wouldn't issue our license until the BATF did and vice versa.

We got our potato barn fixed up with power, water, and cement floors; we had all our equipment ready, and the fruit was ripe in the field. We couldn't hold off the birds much longer, and rot was starting in the Pinot Noir. We couldn't figure out what the government agencies were waiting for. Alex pleaded with them to do a final inspection so we could start picking. BATF inspectors on Long Island had never had a winery to inspect before—they were busy chasing after illegal guns and cigarettes. I sat nervously in Alex's office while he called the agencies and told them that he would hold them responsible if we lost any part of our crop. He went back and forth on the phone between them, negotiating a settlement. At the last possible minute, they all agreed that he could drive into Manhattan to pick up our licenses.

Alex had done all of the business planning and negotiating himself. My job was to set up a lab for testing the wine, and that was fun. I had taken just enough chemistry to read a lab manual, and I loved looking through the lab catalogs at beakers, titration stands, and vacuum pumps. The federal government required us to have an ebulliometer—an archaic-looking French distillation unit for testing alcohol—and a Cash still, which was made of handblown glass and tested volatile acidity, or vinegar. Otherwise they didn't care what kinds of tests we did on our wine, but we wanted to be able to check the juice and wine for sugar, acidity, pH, and sulfur dioxide, as well. Finding labware was easy, but finding winery equipment was not. Alex had to figure out what kind of crusher, press, pumps, tanks, and fittings to buy. Small was bad in this case, too. No small-winery equipment was being made in America. It all had to be imported from Europe, and then all its wiring had to be changed to accommodate the differing power supply. If the instructions that came with the equipment were in English, they were unintelligible. The Ger-

man press we bought, for example, had a big sign that said
CAUTION: DUMP WHENEVER!

For a while, it looked like we wouldn't be able to get the
stainless-steel fermentation tanks we needed in time for harvest.
We hadn't known that they needed to be custom-made. At the
last minute a customer of Mueller Tanks in Missouri couldn't
pay for the tanks that had been made for him, and we were able
to buy them. These tanks had been made with raised manholes,
so that the only way to empty them was to get inside and shovel
them out by hand. Never mind; we were lucky to get them.
When the big tanks arrived on a massive flatbed truck, it was up
to Alex to figure out how to get them past the rafters and onto
their concrete pads inside the potato barn. One tank never did
fit inside. In theory, it should have fit through the large garage-
style door, but it hit the tracks that suspended the door itself.

The first day of harvest dawned with a bustle of activity as we
spread our new, daffodil yellow and grass green plastic picking
boxes along the vineyard rows. These boxes were far superior to
the wooden bushels we had used the year before. Instead of having thin wire handles that would dig into our hands, these had
indented handles built into either end. Each one could hold
about twenty-five pounds of fruit—more than that, and the clusters would start to crush each other. Unlike the bushels, these
boxes could be nested inside one another when they were empty,
or, turned in the opposite direction when they were full, stacked
high on nubs that kept them separated.

In our winery, the converted barn near our house, new equipment stood ready to receive the harvest. We had removed the
original walls of a lean-to section on one side of the old barn so
that it could be used as an open-sided but sheltered pressing
dock. A cart with the stacked boxes of harvested fruit was parked

near this overhang, where a machine called a stemmer-crusher with spinning prongs removed the bitter stems from the clusters as each box was dumped by hand into a hopper. The same machine had rubber-coated rollers that crushed the berries in order to release the flow of juice (called "must") that would go into vats to ferment. There was much discussion about how far apart to set the crusher's rollers so as not to break any of the berries' large seeds. For us, this was a matter of trial and error. We didn't know how the juice should taste in the first place.

From the stemmer-crusher, the must was moved through a powerful pump designed to push the solid berries along with the liquid must. The pump would send this slurry through a long three-inch-wide hose down the stairs into the top of a waiting fermenter, located in what had been the potato storage cellar. All the equipment had been cleaned with a chlorine rinse, followed by fresh, hot water. The new lab, down in what had been the secret bootlegger's cellar, stood ready for me to test the must. Anne, who had learned to walk about a month before, toddled around the scene as the cats wove in and out of the picking boxes. Figuring the Romans would have had a virgin to crush the first fruit of the harvest for good luck, we designated her to give that batch a good stomping. Before the picking started, Alex and I stood on the pressing dock, surveying our new equipment. Once the first cluster went into the stemmer, we'd be committed to the wine.

We had too much fruit to pick by ourselves, but a group of about ten people had read about our vineyard in the paper and volunteered to come for the day to help with the picking. They were full of enthusiasm, and we were thrilled to have them. They had agreed to start picking at eight o'clock, but there was no sign of them then, nor did anyone show up at nine. Alex and I started picking by ourselves. We kept peering over the vines and listening for their cars in the driveway. Shortly after ten, the pickers dribbled in, all in a party mood. They were ready to pick, but first they wanted a tour of the winery. Then they had to find their dark glasses, gloves, sweaters, bottled water, and coffee mugs. When they finally got to picking, it wasn't long before the inevitable berry tossing began with friends taking potshots at each other. By noon they were all hungry. I had been up at five to make them barbecued chicken and apple pie, which they wolfed down gleefully. Then, their bellies full, they worked a desultory half hour more before announcing that it was nap time, and that they had to get back to Queens.

It was with a sick feeling in the pit of my stomach that I waved cheerily at the departing volunteers. Ten people whom we had counted on to work for at least six hours each had just had a lovely outing, and the hours they didn't work now belonged to Alex and me and a few extra people we managed to hire at the last minute.

So who helped the Little Red Hen make the wine? We did it ourselves. Anne played her role, dancing on a bushel of fruit, I ran my lab tests, and Alex orchestrated the crush. We could have used more help, but in fact we were doing what we wanted to do. As for Alex, he made a very good Red Hen.

We did all sorts of things wrong when we made our first batch of wine. All we had to guide us was a fat blue textbook, a companion to *General Viticulture* from the University of California that was more or less a bibliography of winemaking research. "Professor X finds that at 23 degrees Centigrade, upon titration the total acidity (expressed as sulfuric) will not reveal an altered pH due to elevated potassium levels . . ." and stuff like that. The information was virtually useless to us rank amateurs, and singing the Polish anthem wasn't going to get the job done, either. Although we liked to say that we could learn to do what the average French peasant is born knowing, making fine wine is more complicated than that. It's trickier than that episode of *I Love Lucy* where Lucy puts grapes in a big tub and stomps on them. There are decisions to be made at every turn.

Red wine is easier to make than white in the beginning, since the crushed grapes go directly into the tank to ferment, while for white wine the fruit must be pressed to separate the juice from the skins immediately. The first wine we made was red; it was Pinot Noir, and there are more theories of how to deal with Pinot Noir than there are busts of Louis Pasteur in France—but we didn't know that. Use a fermenter that rotates? Heat the must for the first hour? Cold soak it for three days? Add stabilizing tannins? Change the pH with tartaric acid? Put a screen in the tank to keep the skins from drying out as they rise to the top during fermentation? We didn't know that these were options. We did know to keep the red juice circulating so that the color in the skins could percolate throughout the must. We also knew that we had to watch the temperature of the must, since the yeast had to be kept under ninety degrees Fahrenheit to stay alive. But how often should we pump it over and for how long? Should we aerate it or protect it from air? How long should we leave the skins in the tank before we pressed out the new wine? These issues didn't emerge until we were faced with actually doing the job.

Yeast cells are the engines that drive fermentation. They fascinated me just as my colony of ants had held my attention so many years before. I liked to think about how things got done and where power came from. In early photogravures of farms in nineteenth-century America, there are often huge teams of horses pulling equipment designed to tame the wilderness. As larger open spaces were cultivated, greater power was needed to work the evolving machinery—balers, harvesters, and plows—that was a part of the industrial revolution. How wonderful to see agriculture team up with industry. The farmers hitched more and more horses together until eventually one big machine, the steam-generated tractor, took on the power of horses. On the North Fork, in 1888, a farmer whose family had been among the first English settlers in the region invented a potato combine that revolutionized the crops grown in the area. It was a good machine but it was also such a big investment that anyone who owned one had to make a commitment to potatoes. The combine ended a long history of diversified crops in the area, and got farmers to think of themselves as potato farmers just as we thought of ourselves as grape growers. Thus, the mechanization or modernization of agriculture intensified the domination of the farmer over the land, as one crop replaced many.

A vineyard doesn't require massive tractors because the land is plowed only once, at planting, and much of the work is done by hand. Even so, every time I looked at our sixty-horsepower John Deere, I pictured sixty horses pulling the sprayer or the cultivator, and thought about what it would be like if I had to feed, water, and clean up after that many large animals. There had been a stable on our farm near the potato barn that was now our winery.

In choosing to make wine, we were going to have to harness the energy of much smaller beasts than horses. Yeast would do the work of changing the sugar in our fruit to alcohol, and while yeasts are invisible, they still make demands.

Yeasts are single-cell animals that are in the air everywhere. The ripe grape berry is fully designed to become wine; it has a waxy film called subarin on its skin that traps the wild yeasts in the air. There are many kinds of yeasts around, and some of the wild ones will make wine that smells like ripe bananas doused in paint thinner lying in a bed of rotten hay.

In the days before 1870 or so, winemakers in France tried to keep women out of wineries. They couldn't figure out why some barrels of their wines were good, while others turned to vinegar. They did notice that the volume of wine in a barrel rose and fell with the seasons or the moon, and since women were also affected by the moon, logically, women could ruin wine. (See, for reference, the logic in *Monty Python and the Holy Grail*, explaining why a witch is a duck.)

It took a French scientist, Louis Pasteur, to absolve women of causing wine spoilage. In the mid-nineteenth century, the French wine industry was of so much economic importance that in 1863 Napoleon III personally asked Professor Pasteur to direct his studies to solving the problem of chronic wine spoilage. Too much wine was going bad in transit to England, and while the French loathed the English, they still wanted to make money off of them.

Pasteur was a highly controversial man. In the 1850s he had demonstrated that the transformation of grape-juice sugar to alcohol had to be the result of living organisms. Before this, it was accepted that the sugar itself had dynamic properties that spontaneously generated alcohol. People had previously observed yeast, but they believed that yeast was made by the fermentation, not vice versa. A great scientific battle ensued because few people wanted to give up the idea of "spontaneous generation," which was perceived to be the work of God. If they could put a piece of meat in an empty jar and find maggots crawling on it a few days later, didn't that prove that the maggots came spontaneously out of nowhere—that God had put them there? These people thought that babies were made in about the same way. Pasteur was able to prove that invisible microorganisms are ever-present in the air, and that they are responsible for both fermentation and spoilage. He even proved that they could be killed by heating—a process we still call pasteurization. Even so, this understanding of yeast threatened the whole social order— bad wine could no longer be blamed on either God or women.

Besides, winemakers still wanted to keep women out of their wineries lest the wine be tainted by those vessels of original sin.

In 1975, when Alex and I made our first commercial vintage, original sin was not a consideration. Nevertheless, as we began to learn about winemaking, it occurred to me that Alex was like an old-time *maître de chai*. As master of the cellar, he expressed little interest in the science behind the process. He made wine by instinct, whereas I loved to study the molecular permutations of fermentation. We were a good team that way, with me doing the research and him orchestrating the process. "I'm science," I would say, "and Alex is magic."

I liked the science of winemaking, and I was also a mother. The yeast cells were my children and I wanted to understand the physical needs of the microbes that made our wine. At first, I was so afraid of failure that I took a textbook approach. But just as Dr. Spock couldn't be in the room to help stop a tantrum when I was caring for my child, wine professors Amerine and Peynaud weren't handy to discuss the smell of rotten eggs emanating from the fermenter. As with mothering, instinct and intuition proved to be as important as science, if not more so, in winemaking. Making wine on a commercial scale with the number of grape varieties we had turned out to be like having ten children, all needing attention at once.

Because yeast cells are alive, their activity is not purely mechanical. I began to understand that they have a psychology, however primitive it may be. Yeasts are less like children than they are like lusty teenagers whose sole desire is to have sex. The thing is, they don't mate; they just divide themselves. Whether or not it's fun for them, I couldn't say, but it's what they want to do. As long as they have a source of sugar to eat, and sufficient nitrogen and vitamins to create new cells, they keep on dividing with wild abandon.

The sad truth of the matter is that the winemaker's goal—to turn the sugar in fruit into alcohol—inevitably kills the yeast.

As the yeast in its frenzy gobbles sugar to keep up the energy to reproduce itself, it breathes in oxygen and gives off carbon dioxide. In the confines of a fermentation tank, soon the carbon dioxide, which is heavier than oxygen, blankets the tank, and the yeast starts to suffocate. During the few days of vigorous fermentation that followed the harvest, I sometimes wondered if the yeast didn't have it in for me, too. I would walk down the stairs into the wine cellar and feel the suffocating layer of carbon dioxide that spilled out of the fermenters, lying in a heavy, invisible layer on the concrete floor. Alex would be moving hoses and running pumps, unaffected by the gas, while I could stay there only for minutes before I had to climb back out, gasping for oxygen. His height put him above the layer of CO_2, while I, being a foot shorter, was covered by it.

Once the yeast becomes blanketed with CO_2, the only way it can keep up its orgy of reproduction is by switching from respiration, using oxygen, to a different mechanism, fermentation. It's suicide for the yeast, as fermentation creates alcohol. Depending on the strain of yeast and its nutritional status, the last yeast cells may survive just long enough to finish eating all the sugar.

I needed the yeast to make more alcohol, not more yeast, so I was a little like the diabolical mother who asks her children to clean their plates of poison, or the madam who stabs her clients in the back as they lie in bed, exhausted from their efforts. Really, I wasn't that involved with the yeast cells. Maybe they are living creatures, but of a lower order, and anyway, they kill themselves. What I wanted was the wine—deep and rich, or fresh and sprightly. If I had to take advantage of a few billion yeast cells, so be it.

The secret life of wine continues at the microbial level even after the yeast fermentation ends. There are bacteria in wine that are much smaller than yeast. Their needs are more particular, and if they don't find just what they want to consume in the wine, they either die or remain in the wine, impotently, until coming into contact with air or gaining some nutrients when the wine is blended, allowing them to grow. Some bacteria—those that need air—can turn wine to vinegar. Other organisms, called malolactic bacteria, can improve wine when they eat the sharp-tasting malic acid in the wine—the same acid that apples have—and change it to lactic acid, the acid of mellow cheese. If all goes well, this change also creates silky glycerol, buttery diacetyl, and other complex elements that distinguish a great wine from an ordinary one.

Just as there are good and bad yeasts for wine, there are also good and bad malolactic bacteria. The same conditions—warm must with a high nutrient level—encourage them both. The bad ones leave an aromatic hint that something has died in the wine. To make a great wine, a winemaker has to put all at risk, and walk a fine line between complexity and spoilage.

When we first made wine, we would leave the malolactic fermentation to nature, and sometimes the bad bacteria would take over. As soon as cultured bacteria of the good sort became available, we bought them and inoculated the wines rather than leaving the fermentation to chance.

One of the spoilage bacteria that can affect wine under the same conditions that promote the malolactic fermentation is aptly named *pediococcus* or "feet of caca." Virgil didn't tell us what to do about that. Nevertheless, one remedy that we used to improve questionable-smelling wine—carbon fining—was similar to something the ancient Romans used. They apparently had problems that we never had, with larger organisms. Goats or other animals sometimes fell into the pitch-lined pits in the ground that were used for fermenting wine. The Roman practice of hauling out the animal that fell into the fermentation pit, burning it, and throwing the charred remains back into the wine might well have been the first use of activated charcoal—

carbon—to erase foul odors in wine. Fortunately, we didn't have to burn any goats to get activated charcoal to clean up our wines. We could buy it directly from a supply house.

There are other things that can be added to wine, and have been for centuries, in order to "fine" it, or make it better. I experimented in the lab with all of them. The bentonite clay on which we had risked our lives on the Yampa Bench Road became a common tool in the winery. It prevented proteins in the fruit from clouding up the wine and was used routinely in white wines, especially Sauvignon Blanc. The electric charges on either end of the tiny fossils that compose the clay latch onto the opposite charge in proteins. Since the clay is not soluble in wine, it falls to the bottom of the tank, taking the proteins with it.

Ironically, it is the addition of certain proteins that is best for softening a harsh wine. If too much tannin—the same thing that makes tea astringent—has dissolved in the wine from the skins, stems, or seeds, pure protein from blood, egg whites, or gelatins such as isinglass (made from fish) can be added to remove it. I found out in the lab that, yes, excess isinglass smells like fish. However, there are highly refined brands that can smooth out a wine better than anything else.

The biggest drawback I found with fining agents was that they always had to be tested for each batch of wine, because a smidgen too much could strip a wine of flavor, aroma, and color. Still, under the best circumstances they can make the true fruit aromas shine. And sometimes they can be miraculous, as I learned when a leaky pump engine tainted one of our tanks of wine with oil. The remedy? Skim milk. Works like a dream.

🐝

Through the seasons the sun shone and the rains fell, and in 1976 Mother Nature had her way again and I became pregnant with our second child. Once again I enjoyed being pregnant and kept working to the end of my term. A week before our baby was born, I was standing in the produce section of the supermarket with my belly bulging like a watermelon when a woman came up to me and said, "Just don't drop it here."

I held on four days past my due date, as I had with Anne. This time my water broke in the middle of the night. Thinking labor would probably have to be induced as it had been before, I let Alex sleep. He had been up late the night before, moving the contents of one tank of wine to another, after he discovered that the first tank was slowly leaking. What looked like water on the concrete floor had been wine. By the time I decided to wake him up, I was in heavy labor. Leaving Anne with Charlie, we took off down the farm road in our Ford van. Halfway down the drive, I turned to Alex and said, "I think I'd better lie down on the backseat. I'm sitting on our baby's head."

It felt like a basketball. Stretched out in back, I tried to stop the urge to push by panting, until I thought, To heck with it. Why wait? I told Alex that the baby was coming out whether he stopped the car or not. He pulled over next to an open field. A full moon hung low in a lapis lazuli sky. Out the window, I could see the white steeple of a prairie Gothic church, which I took as a blessing for our child.

Alex got to the backseat just in time to catch the next addition to our family. The baby's sweet "Aaah" was a satisfied sound—not a cry, but an announcement that he was pleased to join us on the outside.

Alex gently laid him on my chest and carefully covered the baby with his grandfather's old green cashmere touring blanket. We decided to go on to the hospital, to get us cleaned up and checked out.

After driving along for a few minutes, crowing all the way, Alex realized that we had forgotten to check to see if we had had a boy or a girl. He stopped the car again and made an inspection. "It's a boy!" he announced.

I wanted Alex to know how happy I was that he was this child's father, and how good it felt to have his help in delivering him in this private way. Alex was his own father's namesake but with a different middle name (Mackenzie instead of Davidson) and hadn't wanted to name a son after himself as well. He had suggested Odhar, an ancient family name instead. Was he serious? Maybe. But I thought our son should have his father's name, and now that he could look his son in the eye and see how much he resembled him, right down to the slightly puffy flesh under his chin, Alex agreed that we could call the baby Alexander. We added Stewart as a middle name, so he wouldn't be a Junior. I clinched it by saying, "We can call him Zander for short. You get the first part of the name, and he gets the last."

My sister had had a boyfriend named Zander for a short time, when I was about fifteen and jealous of her success with boys. He gave her an old pair of skis—not the most romantic gift, especially since she didn't ski, but even though I had never met him I thought he sounded ultracool. I didn't explain that to Alex at the time. How could I justify the feeling that having a son with the nickname of my sister's virtually nonexistent boyfriend made everything right in the cosmos? It was just a vague undercurrent left over from my adolescence. What was more important was that our son was alert, strong, and evidently not at all traumatized by having been born.

As a precaution, I spent one night in the hospital with Zander. Because he was considered a "dirty baby," meaning one born outside the hospital, he and I were isolated from the other new babies. The result was that I got my own room, and the nurses let Zander stay with me. This was something I wanted more than anything—to be able to hold my baby right after his birth. When I lifted him out of his cradle, his head rolled around as if it were attached by a ball bearing. I couldn't figure out how to hold him for nursing without that letting little head roll off my arm. It had been nearly three years since Anne's birth and I felt humiliated. "What is wrong with you?" the nurse wanted to know. "How could you forget how to hold a baby?"

After Alex left Zander and me to take the good news back home, I lay in my bed and wept, feeling unaccountably like a betrayer, or betrayed. It must have been due to the rush of hormones that follows childbirth—raw emotion, without subject or object.

The next day, my obstetrician visited me. Sitting on the foot of my bed, he said, "Now you have two children, and I really think that's enough for a farmer." Since my pregnancy and delivery had been so easy, I rather expected him to say congratulations and leave it at that. I wasn't planning to have any more children anyway. Back then, we were concerned about the population explosion and had agreed from the beginning of our marriage not to exceed our self-replicating allotment of two children.

One neighbor said to us, "Now you have millionaire's children." When you have a girl and a boy, he explained, you need to be a millionaire because you can't hand down clothing from one to the other. Nice idea, I thought. I still have photos of Zander wearing Anne's handed-down flowered overalls. Although she was only three, Anne somehow figured out that girls were supposed to dress differently from boys. She herself liked to wear frilly hand-me-downs from her cousins. One day shortly after Zander was born, Anne decided that she could use the difference between girls' and boys' dress to keep me from going out in the field to work. She tried to stop me from putting on my jeans. Grabbing at my pants leg, she cried, "Mommy, wear a dress!" It was a good ploy, but it failed. I had hired a part-time baby-sitter to watch Anne while I did field work. There I was, disappearing just like my father had. To Anne I might just as well have been in Manhattan.

The summer that Zander was born, besides welcoming Charlie back onto the farm, we also hired a man named David to help us. David had learned all about seafood while working with his father, who had an aquaculture business in South America. He thought he might like to learn about wine, too, but he spent only a few weeks working for us before his father called him back to South America. I remember David because he introduced us to eels. He had been delighted to discover that the creek near our house was full of them, and he gathered a basket of them to cook on the grill for us and our friends. I had never eaten them before, but I was game. I was adventurous.

David took an eel out of the basket. He nailed its head to the maple tree in our yard. Cutting a little circle around the head, he pulled the skin of the eel down the length of its body, inside out. I thought it was like taking a sweater off a small child. Someone else said it was like a circumcision. We grilled the eels and ate them. Pretty good, I thought.

After we ate, while David was cleaning up in the yard, I went inside to get Zander up from his nap. As I passed by the chimney, I saw a piece of rope about a foot long lying on the old linoleum. Absentmindedly, I bent over and picked it up. It was cold, and then it undulated in my hand. The piece of rope was a snake. It had come up the chimney from the basement and was resting on the floor. Screaming, I threw the snake back on the floor, ran out of the house, and said, "David, get the snake! You have to get the snake! Get it out of there!" Looking nonplussed, David brought out the little snake. He said, "What's the problem? It's a garter snake."

The problem wasn't that it was a snake. The problem was that it wasn't what it appeared to be.

There wasn't really any dangerous wildlife on the North Fork as far as I could tell. I thought our little yard was completely safe.

With Zander on my back now, Anne played contentedly outside, making mud pies and squishing fruit from the weeping mulberry bush. She was a sunny three-year-old with chubby legs and "boing-boing" curls. Anne had always seemed content wherever she was, sitting on a blanket at the beach as an infant, or at the table where she drew interesting squiggles for long periods of time. One summer evening while I was preparing supper, she went into the yard by herself. I could see her fiddling about from the window. I took my eye off her for a few moments— obviously, too many—for when I looked for her again she had disappeared. I ran outside, calling her name, but she was nowhere to be found. When I alerted Alex, he began charging around the same way. There were no limits to our farm—no fences or walls all the way to the road. Besides that, there were three different dirt roads that led to our yard. We couldn't figure out which way to go to look for her. Alex got in the car anyway and took off down the broadest driveway while I walked into the vineyard. The foliage was dense, and the vines had grown to the top of the trellis. Soon the sun would set. It was a desperate situation.

In my frantic state it took me several awful minutes to notice that there was music coming from a house on the other side of the woods. They must have been having a party. And there, amid the vines, was little Anne, dancing happily. Her arms were raised; her skirt was twirling.

Not long after that Alex told me he had heard that a neighbor's child had been attacked by feral dogs. We lived near the town dump, and every year the dogs abandoned by people at the end of their summer vacations would find their way to the dump looking for food. After a season of living at the dump, the dogs would band together in packs and roam around town, looking for fresher fare. I never saw them myself, and I didn't quite believe that dogs could really be that dangerous in this sleepy little village, but I did agree to keep the kids inside for the time being. Missing her backyard freedom, Anne would protest, "I wanna go *somewhere!*"

Shortly thereafter, I heard noises in the field behind our house, and saw a group of men armed with shotguns, stalking the wild dogs. Their gunshots rang across the meadow. I guess they got the dogs, but I later wondered how many other wild dogs were out there.

Barely a month after Zander was born our winemaking efforts also came to fruition when we released our first wine. It was made from fruit picked in 1975 and aged in barrels until we bottled and released it in July 1977. When we harvested the fruit, we hoped that we had made one of the world's great Cabernet Sauvignons, and so we followed the French practice of keeping it in cask for two years. Our first barrels came from a cooper who made whiskey barrels from Arkansas oak.

We had had the cooper send us samples of the kinds of wood that he could use to make our barrels, and we selected what we liked best. I poured boiling water over shavings from these staves, and when the water cooled, we tasted it. There were often people wandering into the vineyard to see what was going on, even before we had wine for sale, and we had some fun with one Manhattan couple who dropped in just as we were doing our oak tasting. They professed to be great wine aficionados, so with ceremony, we poured samples of the oak-flavored water into large wineglasses, swirling and sniffing them as if they were wine. The tannins from the wood had colored the water so that it actually looked like white wine. So misleading was the nutty, smoky aroma of the wood that the couple declared that the "wine" was as good as a French Chablis.

Although the aroma of these staves was appealing, the effect of putting our rather light Cabernet into the barrels that were made from them was disastrous. After several months the coarsely grained American oak gave the wine a sort of turpentine aroma. The charred inner surface of the barrels stole the color out of the wine, too. In the end, it smelled more like Scotch than like Bordeaux. When it came time to sell it a year and a half later, we decided to face reality and market it as a nonvintage rosé.

Imagine showing the world what Long Island could do with a wine that had a pale orange color, an aroma of the heath, and an astringent, bitter finish. Believe me, we were in denial, and so were our forgiving customers. At six o'clock in the morning on the first day we released this wine, there were people camped on our lawn as if it were a yard sale. State law prohibited us from selling wine until nine o'clock, but as soon as the hour hand passed nine, we sold our first bottles. We made so little of this wine that we had a "no tasting" policy. We said we couldn't give away something so precious, but in reality, we would have sold considerably less wine if people had been able to taste it first.

7

Survival of the Fittest

EVEN WHEN WE TURNED OUR ATTENTION TO WINEMAKING techniques, our focus was in the vineyard. The whole point of our wine was based on our ability to grow premium grapes on Long Island. We thought the vines were doing well, too. They gave us a reliable crop every year, and the fruit was good. In 1979, shortly after our harvest, Dr. Shaulis brought Dr. Kasimatis from California back to see our progress. They came down the road in a cloud of dust and we gathered in a little group, full of excitement and ready to show off what we had done. Dr. Kasimatis was all smiles as we ambled into the Cabernet Sauvignon. This was a section of the first vines we had planted. It was also the location of Dr. Shaulis's pruning trial. As it turned out, the untested rootstock that this Cabernet had been grafted to, a hybrid called Baco Noir, had proved to be excessively vigorous, so that the vines' growth had been extravagant. Dr. Kasimatis was entertained by our description of what it was like to drive a tractor down the rows that had been trained in the Geneva Double Curtain. These vines were suspended at six feet on T-shaped arms so that they hung over the alleyway by a couple of feet on either side. To cultivate or spray that section at normal speed was to risk decapitation. Fortunately, there were only a couple of rows like that. We explained to the professors that, since they hadn't produced better fruit than the other training systems, we were planning to remove the GDC in the winter. Shaulis winced as we demonstrated the limitations of his pet trellis, but he could see, if he opened his eyes, what we were talking about.

We went on across an alleyway between the Cabernet and the Pinot Noir. Shaulis pointed out that some of the leaves on our plants had turned various shades of red. "Well, it's October," said Kasimatis. "Some color change is to be expected as the vines senesce. There could be a touch of virus here also. Didn't these plants come from uncertified stock upstate?" he asked, turning to Alex.

"It's true," Alex replied. "These plants are uncertified. But all our newer plantings, like those Merlot at the foot of the hill, are California-certified virus-free stock."

Shaulis and Kasimatis glanced down the hill. Kasimatis got agitated. "Just look at them," he exclaimed. "Those Merlot have entirely red leaves. They have classic symptoms of virus."

Alex and I gave each other a dire glance. How could the certified virus-free vines that we had been forced to buy from California have a virus? But Kasimatis was insistent—they were riddled with it. They had crown gall, too. That much we already knew. Many of their trunks had split open as misshapen growths disfigured them. This had started the first year they were planted.

After tending those vines for five years—five years of sweat and toil, not to mention expense—we had to pull them out. In the process of investigating the situation, we were told by an insider that the researchers in charge of the certification program had known that this clone of Merlot had a virus, but it was a mild virus. As there was no other virus-free clone that they could sell, they decided to certify it anyway. We sued the nursery that had sold it to us.

The lawsuit lasted several years, as these things do. Our attorney managed to set a precedent for the case to be heard in New York courts, since that was where the injury occurred. It didn't matter, because the nursery went bankrupt before a trial could start. In the end, we settled for the initial purchase price of the plants, about eleven thousand dollars. We never did replant that part of the vineyard. Within ten years of planting the certified vines we had bought in 1974, we had pulled all of them out because of disease problems that had come with them.

Despite the troubles we had had with poor plant material, we were starting to make an impression on the wine world, and several other people had followed our example in planting wine grapes. Besides Dave Mudd, who never made wine but who continued to install vineyards for other people, there were a couple who traded their cutting-edge restaurant—the first to serve rose petal sorbet in Westhampton Beach—for a new vineyard; a jingle writer from New York; and a physician whose other investments included nursing homes. We were happy that all of the new growers had planted vinifera, and that they all planned to make château-style wine, or wine exclusively from their own estates. Anyone could have bought a site on Long Island and used it to bottle cheap wine from another region, just to take advantage of the New York market, but these newcomers didn't do that, at least not right away. Dave organized them, as well as anyone else on Long Island who grew so much as a single grape, into the Long Island Grape Growers Association.

In the autumn of 1983 the association had a family picnic that took place on a grassy slope, overlooking our vines. While parents sat on blankets and drank wine, the children played in the meadow. Zander, who was six, found a ribbon snake. It was black with yellow stripes. All the children were excited by it. I was still embarrassed about my reaction to the snake that I had thought was a rope years before, so I made a point of picking up this snake very casually and encouraged the children to hold it. When Zander said he wanted to keep the snake, I couldn't say no. He called it Freddy. So Freddy came to live with us in a box in the kitchen. After a week or so, I started to worry about Freddy. Freddy must be hungry, I thought. But what do snakes eat?

I didn't know, and I didn't have time to go to the library to find out. Probably insects, I thought. I asked Alex. "Probably insects," he said. We conferred about what kind of insects the snake would like to eat. Since it was October, there were plenty of crickets that were moving into the house to get away from the impending winter. "That should be a good dinner for Freddy," I said.

I was a bit squeamish about feeding a cricket to Freddy. I knew that in China, crickets are symbolic of a happy home. I had seen a collection of intricate cricket houses at the Kansas City Museum, when we had stopped there on our way back east, in the summer of 1972.

After Freddy rejected the lettuce we offered him we decided to go ahead with the cricket. It seemed like it would be a good lesson for the kids about the food chain: every creature has to eat something, and the food of choice is often another animal. The cricket was released into Freddy's box. It sized up the snake. The snake sized up the cricket. The cricket jumped onto Freddy's head and ate his eyes. Then it ate his brain. So much for nature lessons.

Without getting misty eyed about it, I tried to live in peace with all the insects I encountered on the farm. Not that I wouldn't swat a fly, smash a mosquito, or pinch a spider if I found it in my bed, but I shared Alex's "you don't hurt me and I won't hurt you" approach to most bugs. When it came to crickets, though, I had a hard time being sympathetic to them, especially after seeing what happened to Freddy. When they came into the house every fall, sawing away at their fiddler tunes, I debated whether it was worse to crunch their little bodies with my boot or to use the vacuum cleaner on them. Then I'd feel guilty and leave them alone.

There was another insect prevalent on Long Island for which I had absolutely no sympathy: the primitive and sinister tick. I was familiar with ticks even before I worked outside all day on the farm. When I was a child I would sometimes find a tick in my hair after I had jumped in a pile of leaves or walked barefoot through the meadow. Ticks burrow into your skin and suck your blood. You can't pull them out without taking a bit of flesh with them and leaving a scab. They look like round black dots with tiny legs until they get so bloated with blood that they look like inflated gray balloons. I have a distant memory from when I was about four of sitting in a warm bath one summer evening, enjoying the soothing water, when my nearsighted baby-sitter noticed a mole that I had on my inner thigh. Insisting that it was a tick, she tried to remove it by pinching, plucking, and scrubbing it. I screamed bloody murder as she tightened her grip on me and scrubbed all the harder. Finally, she came to the right conclusion and stopped.

Besides that, my cousin John had nearly died from a case of Rocky Mountain spotted fever brought on by a tick bite. But out in the vineyard, I rarely got bitten by ticks, and I didn't worry about them. There were only so many things I could worry about at once, and even though I detested ticks, they weren't on my list of worries.

All our pets brought in ticks on their fur. Alex took on the task of removing them carefully with tweezers so he wouldn't have to touch their engorged bodies. He drowned them in a jar of alcohol—a sanitary demise that was preferable to burning them.

One day when I came in from the field, I found him lying in bed with a high fever. He had an intense headache and within a few hours became delirious. Other than a strained back and a finger that was almost severed when a glass carboy broke as he was moving it, Alex had not been sick since his back surgery, so he didn't have a regular doctor. I took him to the surgeon who had stitched up his finger, an old-timer with an "I'm God; who are you?" approach to medicine. The doctor wanted to do a spinal tap. Alex had had a spinal tap before his back surgery and that was an experience he would never willingly endure again. Delirious as he was, he refused treatment. The doctor was furious and told us to leave, saying that he would not bear any responsibility for the outcome of Alex's illness. Alex went home, loaded up on aspirin, and, with all the determination he could muster, got better.

I thought that was the end of it, until he had another episode much like the first one. It happened again several months later, and then every so often over a period of about five years. Sometimes he would get burning hot, and sometimes he would get freezing cold. One day he got so hypothermic that I piled comforters on top of him and then gathered the children to lie on him with me to warm him up as he shivered violently beneath us.

Besides the fevers, the headaches came on even more frequently. Then he became fatigued, lying in bed or sprawling on the sofa to rest whenever he wasn't needed in the office or cellar. By then he seldom worked in the field. When duty called he would rally himself and go into the world with apparent energy and strength, so that no one outside the family knew he wasn't at the top of his game.

Alex did not want to go back to a doctor just to hear that he needed a spinal tap, but somewhere along the way he confided in a friend. Dr. Tom Cottrell was a neighbor and a pathologist at the regional hospital. He commuted an hour to his job so that he could live in Cutchogue and sail in Peconic Bay. He had recruited Alex to serve as his crew in the regular Wednesday-night boat races around Robin's Island in Peconic Bay. In the course of their sailing, Alex came to trust Tom. He learned that Tom was also a dean at the hospital and knew all the doctors there, so he asked him to recommend a good diagnostician. Tom sent him to Dr. Green.

Alex liked Dr. Green immediately. He was an internist who specialized in cardiology. The man worked out of an office in his comfortable house in Patchogue. His wife, Sue, a pleasant and attractive woman, was also a nurse who ran his office and assisted him. Dr. Green arranged for Alex to have a number of tests, including a CAT scan. He feared that Alex might have a brain tumor, but there was no way of knowing until he ran the tests. With pain located at the base of the skull, as Alex had, there weren't many possible explanations besides a tumor, the doctor said.

Back home, we braced ourselves for the worst. When the tests came back showing a healthy, tumor-free brain, whatever rejoicing we felt was tempered by the fact that the headaches remained. Without a diagnosis, there was nothing for Alex to do but bear it.

Not long afterward, Tom Cottrell called us with the news that Dr. Green had murdered his wife, that nice nurse who shared his life and his work. No wonder Dr. Green hadn't been able to focus on Alex's diagnosis. He had been having an affair with the church organist. While everyone thought he was in Washington for a medical conference, he had turned around at the airport, flown back to Long Island, and murdered his wife. He had slipped up, though, and as soon as the police discovered his premature return the jig was up.

So now Alex had consulted one doctor who wanted to do a

spinal tap on him and another who was a murderer. There would be no more doctor visits until one fine day when Alex read an article in the *New York Times* about a newly identified disease spread by ticks called Lyme disease. The article included case histories, one of which was exactly like his own.

Lyme disease is carried by a pinhead-sized tick called a deer tick. The populations of deer in our area, and around Lyme, Connecticut, where the disease was identified, had greatly increased in recent years, as the local deer had no natural predators. Deer ticks love venison, but they love mouse blood just as much. As the land became populated the common white-legged mouse had fewer and fewer predators, too. We had been mouse-free while we had cats, but by this time all our cats were gone. These pointy-nosed reminders of the wildlife that still surrounded us enjoyed the habitat of our prerevolutionary cellar. We would set out poison down there, but it wasn't as good as a cat because a cat will eat the mice it kills, while the poison left us searching for stinking, rotting dead mice that had succumbed to it.

There was no way of knowing whether Alex had been bitten by a tick that had feasted on deer or on mouse, but whichever it was, his symptoms indicated that he had Lyme disease. He called Tom Cottrell, who immediately arranged for Alex to become part of a study using an experimental drug supplied by the Centers for Disease Control. Apparently Lyme disease is caused by unusual organisms called spirochetes. These are similar to the organisms that cause syphilis. Unlike most microbes that are prevented from crossing the blood-brain barrier by the membrane surrounding the brain, these spirochetes can enter the brain and cause extensive damage. Most drugs can't reach them, on account of the barrier, but this experimental drug looked promising. To be effective, it had to be injected in the buttocks once a day for five weeks.

Tom Cottrell came to our house on his way to or from work every day to give Alex his injection. For the only two days he couldn't come, he taught me how to do it. Alex went around with a sore behind and a happy countenance as the headaches and fatigue subsided over the next few months.

8

Pet Fever

WHILE ALEX'S BATTLE WITH LYME DISEASE PUT A STRAIN ON the family, a new addition to our household created much joy. Around the time Zander reached his seventh year, Alex announced, "A boy needs a dog." He must have been thinking of the good times he had had as a boy with his Labrador retriever, Abe Lincoln. I had met Abe before I married Alex. He was the sort of genial animal that would lope alongside kids on bikes and wait at the door when school got out. That was the appealing part. He was also known to take off into the hinterlands for days at a time. Someone claimed to have seen him in the Bristol Hills some thirty miles from Alex's home in Rochester.

I had grown up with a dog, too. I had been fond of our dog, Jester, but I didn't like the way he was always begging for attention with his dog breath in my face. The smell of dog food disgusted me, too. And Jester, like Abe, would disappear from time to time, setting in motion a family-wide panic with frantic phone calls to the neighbors. I wasn't sure that getting a dog was a good idea.

"We'll get the dog neutered," Alex promised, "so he won't roam. And I will feed him."

Then the kids got involved. "Please, oh please. We'll take care of him, we promise," they said, jumping around excitedly.

I caved in when Alex suggested that we find a standard poodle. My grandparents had always had standards, and I knew them to be the most intelligent and discreet dogs. Discounting their frivolous looks, I liked the way they pranced along with an air of authority.

We found a breeder up the island with a six-week-old litter of black standards. Her bitch was named Jezebel, and at first I thought, Do I want a bitch with loose morals to be the mother of our pet? The father was a champion dog named Black Dragon, and that sounded better. We found the puppies in a tidy enclosure in the owner's dining room. There were five balls of black fur all scuffling with each other. A sixth pup stood and watched. We took a few of them out of the pen, and the brawling continued with all but the stand-alone puppy. He went off on his own for a stroll under the dining room table. Noticing us, he turned around and marched right up to us, looking us over with his soft, beady eyes. "That's the one," Zander said, picking him up for a cuddle. "His name is Zeus."

I went back into the kitchen with the owner to discuss the care and feeding of puppies, and I started to wonder if there wasn't something wrong with this puppy. The vet had warned me against choosing a runt. This pup wasn't small, but he was almost too quiet. Could he be a bit retarded? We started to reconsider some of the other dogs. Then the door flew open and the owner's son appeared. Zeus went bounding up to him, tail wagging and eyes gleaming. The dog was not retarded.

When I took him to the vet for a checkup the next day, Zeus passed with high marks. But when I told the vet that I had gotten the dog for my son, he said, "Good luck. Don't count on a seven-year-old to take care of a dog. This dog is yours." The vet was right.

I still wasn't sure I wanted a dog, but now that I had one I made up my mind that at the very least Zeus would be a dog with good manners. I got a dog obedience book that advocated using a training collar along with firm commands, and I made it clear to the family that Zeus was never to be allowed on the furniture or in a bedroom (in case of fleas). He was also never to be fed scraps while we were at the table.

Zeus was eager to please, but only to a point. He and I had a major standoff when I tried to get him to sit. I would say, "Sit!" very firmly, and push his rear end down. He would sit, then pop up again, all the while averting his gaze from mine. This happened at least fifteen times in a row. Disgusted, I was about to give up and let him have his way. I'll try once more, I decided. I looked him squarely in the eyes and commanded, "Sit!" He sat, and he stayed.

After that Zeus became my constant companion. Of course Anne and Zander played with him, but often their play was rough, and he would get too excited, bounding in front of them so that they tripped over him, or nipping at their shorts.

Every day, Zeus would wait for me on his bed by the front door. When I made a move to get my vineyard supplies, he would start bouncing up and down at the door. If I said, "Do you want to come?" he would become nearly frantic. As I walked out into the vines, he would frolic ahead, zooming left and right to check out the territory. He would sniff, dig, and chase in the vineyard for about forty minutes. Then, he would take a position under the trellis, near where I was working, but not intrusively.

Every once in a while something would catch his eye and mobilize him again. One afternoon when I was working late I saw him leap up and tear off in his bounding stride. He had spotted a fox. The fox led him on a chase all around the vineyard, zigging and zagging under the wires and into the hollow. I saw Zeus's pace lag, and then I saw him stop, his chest heaving. The fox dashed on for a few more paces and then turned toward Zeus. It gave Zeus a mocking glance, as if to say, "What would ever make you think that a poodle could outrun a fox?"

When it was time for lunch, at noon, Zeus would anticipate it by five minutes, approaching me from behind and signaling the time by poking his snout into my hand. Zeus was more accurate than a watch. And he didn't get fed at lunchtime; he just wanted to let me know it was time to go inside for *my* lunch.

Alex was the dog chef. His *specialité de la maison* was dog gravy that he made in a big measuring cup, a week's worth at a time. Sometimes he added bits of cheap canned pâté to it. Aside from his ears' falling in the gravy, Zeus's table manners were impeccable. He ate his own dinner from a dog bowl in the kitchen while I was cooking for the family. When we sat down to eat, he would park himself on the rug well behind my chair. He appeared to be snoozing peacefully, but he knew exactly when I was finished eating. As soon as I placed my knife and fork together on my plate, he would get up slowly, in a nonchalant sort of way, and position himself by my right arm. Even if I wasn't paying any attention to him, he knew that pretty soon my arm would touch his fluffy head and I would pet him.

Anytime I went out in the car, Zeus came along for the ride. He sat in the backseat, but while I was off doing my errands, he would move up into the driver's seat. There he would stay, with his nose held high, looking just like a supercilious chauffeur. As soon as I returned, he went back to being a dog again.

Zeus, with his gentle manners, became a prime attraction for visitors to the winery. He never jumped up on anyone or barked unnecessarily. With people he liked, he would observe them for a while and then lean against them carefully as a gesture of approval. In fact, he would sometimes become so relaxed that he would actually fall down as you scratched under his chin. We used him as part of our interview process. If we were interviewing people to help out in the tasting room and Zeus didn't lean against them, we thought twice about hiring them.

Other than his interviewing skills, I thought that Zeus wasn't much of a watchdog until the day came when an old drunkard wobbled down our driveway on a beat-up bike. He wasn't a casualty of our tasting room, just an old farm worker who had taken a wrong turn. Zeus let loose a ferocious and truly terrifying howl with bared teeth and a ready-to-attack crouch. The drunkard hightailed it in a hurry. The only other time I saw him posture this way was when a woman wearing a coat with a fur-lined hood arrived at the winery. I guess Zeus couldn't figure out what sort of beast she was—and, as luck would have it, she was mortally afraid of dogs. After he bared his teeth and howled at her, I had to put poor Zeus in the back room until she left.

When we took him to a picnic on the beach in East Hampton, as soon as I jumped into the surf he became desperately worried, leaping and barking along the shore. It upset him just as much when I went on a high swing in the playground. To get back at me for scaring him he took a flying jump over the long picnic table that was set up with corn on the cob and lobster. Anne had taught him to jump like a horse, and he could easily clear four feet. That was the end of Hamptons picnics for him.

Everyone paid attention to Zeus, but it was clear that he was thoroughly attached to me and I to him. Anytime I was upset I would take him for a walk in the vineyard, or have a little tête-à-tête—or nose-to-nose—with him. He would make low, guttural sounds and I would make them back. One afternoon, as Zeus lay in his usual place behind my chair, Anne and I sat chatting at the kitchen table. Anne remarked with conviction, "You love that dog more than you love me."

"Of course I don't," I replied. "He's just a dog."

Zeus stood up as if he had the weight of the world on his shoulders. Without a glance he slouched out of the room and sank onto his bed in the hall with a loud sigh, facing the wall and refusing to look at me. Despite my fervent apologies and avowals of deep affection, he refused to look at me for a week.

I don't know if it was pet fever or what, but at about the same time we got Zeus for Zander, we got a horse for Anne. Boy gets dog; girl gets horse. It made the farm more complete. On the other hand, when I told my mother we were getting a horse, she guffawed and said, "*You? A horse?* Every horse you've ever looked at has run away with you."

My mother was right about that. If I went nose to nose with a horse of any size, that horse would invariably put its ears back and snort goo all over me. I had taken riding lessons, following my sister's joyous example, and it was true: the same horses that went docilely around the paddock with Wendy would rear to throw me off. If that wasn't enough, they'd try to rub me off on a tree or gallop back to the barn. I had no authority over horses. And that rocking motion that takes a horse from a trot to a canter would always send me flying between the horse's ears.

Nevertheless, I retained the fantasy of being like Laura Ingalls Wilder, riding over the high prairie grass feeling wild and free. If I wasn't going to have that experience, maybe my daughter could.

Alex had no interest in riding—it wouldn't have been good for his back—but when I promoted the idea of getting a pony he went along with it. Neither of us knew how quickly and deeply in over our heads we would get with horses—or how much of a strain it would cause in our family. It wasn't that we fought about it, because both of us could see how good it was for Anne, but it was a drain on our finances that we could ill afford. Besides that, it was a drain on my energy and attention. Of course, that's probably why I liked it—it was a diversion from the twenty-four-hour-a-day focus on wine. It was delightful to walk to the paddock as the sun was coming up and have a nickering horse with ears perked forward come to greet me, or to run my hand with a brush over a horse's warm flanks on a freezing day.

I doubt that I would have had the nerve to buy a horse were it not for a wonderful young woman named Denise who worked for us at the time. She had a degree in agriculture and experience with all sorts of farm animals. Denise had grown up with back-yard ponies and knew how to take care of them. Through an ad in the local newspaper she found us an aging but well-trained black quarter horse named Silhouette. Denise gave Anne riding lessons and helped me set up the stable and paddock.

Zeus was furious about Silhouette. If I got on her for a little trail ride, he would follow closely behind us, barking and nip-ping at Silhouette's heels. She would try to turn around and nip him back, or just take off like all the other horses I'd ridden and try to lose me out in the field. Fortunately, he let Denise and Anne ride her without harassing them. Anne enjoyed riding, but she was a timid rider and too small to exercise control over this very wise horse. When Denise left us to have a baby, I no longer had a way to train Silhouette. I ended up sending her to a nearby stable, where the trainer could keep her fit and give Anne lessons. This is when things started to get out of hand. From having a backyard pony that could graze in a paddock, we were suddenly spending hundreds of dollars on board, vets, lessons, and equipment. It got worse from there, because all the kids at the barn were expected to show. Now we had show fees, extra lessons, and eventually the cost of a trailer to haul the horse around.

In the beginning I was opposed to the show business. I knew how notoriously political horse shows were, and I wasn't inter-ested in the competitive side of it anyway. What I hadn't figured on was that Anne loved to show. It gave a focus to her riding. She was a child whose mind was often in the clouds, a day-dreamer, and riding brought her back to reality. Anyone could see a difference in the way she now sat up straight and made an effort to organize herself. It gave her a social life, too, for kids of all ages at the barn worked and played together in a nice way.

Little Zander, aged seven when we got Silhouette, at first was just as excited as Anne about riding. While we still had her on our farm, he'd get on her after Anne had finished riding and walk her around the paddock. Then one day he got a little too gung-ho in the way he urged the horse along. I saw him dig into Silhouette's belly with his boots at the same time he pulled back on the reins. Trying to resist the bit in her mouth, she lunged forward and then ran into the field, jumping over a bale of hay that was in the vineyard to control erosion. Zander tumbled to the ground and decided that that was enough riding for him. He preferred his land-based sports. One thing, he told me much later, especially bothered him about the whole riding scene. He couldn't understand why one stable where Anne took lessons was called Strawberry Fields. He knew the words to the Beatles' song, but he didn't know that Strawberry Fields is an English insane asylum. It made no sense to him that they sang, "Let me take you down, 'cause I'm going to Strawberry Fields. Nothing is real." Strawberry Fields *was* real. His sister rode there. But, he said, "there were no strawberries. I wanted strawberries, but there was only horse poo."

We had had Silhouette for about a year when I got a call from the stable where she was boarded. "I'm afraid that Silhouette has colic," said the trainer. "The grooms found her this morning, writhing in the paddock. I've called the vet, but it doesn't look good."

Horses have long intestines that easily get blocked or twisted in knots. They have no way of burping or regurgitating if they have gas, and colic is life-threatening for them. Anne, Zander, and I rushed to the barn to be with Silhouette when the vet came. It was a sorry sight—the poor horse stood in one place, her head down, pawing the ground uncomfortably. The vet gave her whatever vets administer for colic, but he said we would just have to wait and see if Silhouette could recover from her distended gut. Anne stayed at the barn in order to lead Silhouette around, with the idea that the motion of walking might readjust her innards.

By midafternoon, Silhouette was looking better. We decided to go ahead and visit my parents up the island as we had previously planned. The next morning, we called the stable and heard that Silhouette was looking even better. On the way home we stopped there to see her. As we rounded the paddock, I saw a black heap lying in the courtyard, with stiff legs extended toward the gate. It was Silhouette's body.

Torrents of tears and a big bill for a backhoe rental later, we agreed to start looking for a new horse. We could have just used the backhoe to bury several thousand dollars instead of the dead horse, and saved ourselves a huge amount of pain. But then, five horses and several trainers later, Anne would not have become a skilled equestrienne or a disciplined student, and I would not have learned how to braid horses' manes, muck stalls, or drive a trailer.

During the same time that we got Zeus and Silhouette, we had the idea that we could raise sheep. We had had the vineyard for ten years, and the weeds were still a problem. Every spring they would look really small, and the next time we turned around, they would be out of control. We used weed sprays, mowers, hydraulic sidehoes, cultivators, and, always, the hand hoe. I thought we should get after them earlier, but in the springtime there always seemed to be other things that had to be done first. The vineyard couldn't be cultivated until the vines were pruned and tied. I was in charge of the tying crew, and since we had to work barehanded, we couldn't tie until the temperatures were above fifty. We were also delayed by spring storms. The pruning wood had to be pushed out of the vineyard, and all the wires had to be tightened. I wanted to hire more people to help out, but wine sales were slow in the spring, and Alex feared we couldn't afford it. The sheep were his idea of how to control the weeds. At the same time, they would dot the landscape to complete the pastoral ideal. And they were multipurpose; they would make the perfect accompaniment for our Pinot Noir—lamb.

With great excitement we got piles of literature on raising sheep. It looked pretty easy—just make sure they have water and worm them occasionally, and they will graze their way into becoming chops. Denise helped us select a breed, and we bought two lambs named Phyllis and Lucy. They were adorable, bleating little bundles. We set them free in the vineyard, and they immediately ate all our rosebushes. There went the Chrysler Imperials, and the Peace roses. Weeds? Not interested. As the lambs grew older and larger they became more aggressive. We added two rams to the herd and penned them in with the horse. They liked the horse, but they went after Anne and Zander at watering time. Their hindquarters became matted with manure. Alex found out that the Scots call the little clumps of dung that cling to a sheep's rear *nap daglish,* or something that sounds like that—it's not in my Scottish dictionary. If a Scotsman wants to tell you to get going, he'll say "Rattle your dags." As far as I could tell, that phrase was the only useful thing that came of those sheep. We gave the rams to a petting zoo, and though we did turn Phyllis and Lucy into chops, it was hard to eat those critters no matter how pesky they had been. We gave most of the meat away. The great four-legged lawn-care scheme was over.

9

Growing Up

ALL OF OUR ANIMALS, TAME AND OTHERWISE, WERE JUST A DI-
version. Work in the vineyard and winery ruled the day, every
day. We hired anywhere from two to twenty people, and when
we chattered constantly it helped us stay in motion as we moved
down the vineyard rows or stood on the bottling line. None of us
had grown up doing this kind of work, and we needed some social contact for motivation.

In the vineyard's early days, we would often play games with
one another as we worked. We played "I packed my grandmother's trunk, and into it I put . . ." a list of alphabetical items
that grew as we took turns reciting and adding to the list. We
made up our own vineyard anthem, a sort of sea chantey about
the vines. I regarded the people who worked for us more as personal friends than as employees. We worked so much of the time
that there was little chance to make friends outside the vineyard.

We teased each other and goofed around for the first five years
or so. It made the time go by like recess in elementary school.
One day in March 1975, Alex and I were doing the pruning with
the help of a muscular young surfer named Wayne, when Wayne
made a bet with us. The wind was howling and the vineyard
floor was a morass of frosted mud. I don't know why he thought

of it, but out of the blue Wayne bet Alex and me a jug of wine that I couldn't carry Alex from one end of the vineyard to the other. Alex was six-six and weighed 170; I was five-five and weighed 110, so it looked like a good bet on Wayne's part. What he didn't know was that before we started the vineyard Alex and I had gotten into shape for backpacking by carrying each other piggyback. That had been a few years back, but I was pretty sure that I could still do it, because I knew I was even stronger now. The key was that Alex's legs were so long he could wrap them around me and lock his feet together, making himself into a tidy package for me to carry. The hard part was getting in motion. I managed to hoist Alex onto my back and start walking, but when I got to an alleyway, Wayne began to argue about how far I was supposed to go. I had to stop while we discussed the terms of the bet—with Alex on my back all the while. Finally I said, "To heck with it; I can go the whole way." Pushing myself forward, I made it to the hedge by the house and unlatched Alex's legs from my waist. When I took another step forward, unburdened, I felt weightless. Alex and I both went dancing and spinning around, hooting and hollering. It didn't matter that Wayne never delivered on his bet.

Once the work of the vineyard became more routine the games stopped, although the conversations were still sociable. Most of the time we talked about food. We were all hungry, all the time. For a while I made lunches for myself and the crew when we came in from the field. They would hang out in my kitchen, yakking, while I made sandwiches. In the winter I usually made grilled cheeses to warm us up. There I would be, slicing bread and cheese while my children tugged at my legs and the rest of the crew leaned back in kitchen chairs, joking and relaxing. There were muddy boots scattered all over the floor, mixing with the mess of toys that I never quite got around to picking up. I cooked the sandwiches individually in the toaster oven, and by the time I had cooked my own sandwich, the lunch hour would be over. One day the toaster oven broke down before I got to toast my sandwich. It was as if I broke down with the toaster—I couldn't stand being the one who was still working while everyone else was having fun. "That's it!" I said. "From now on, everyone brings his own lunch."

That changed the dynamic with the crew. Once I stopped feeding them, they started going to the deli or eating in their cars at lunchtime. We were all still friendly, but it wasn't the same.

Back out in the field, whenever there was a lull in conversation, I continued to think about my children or plan the evening meal. I hate to admit it, but often I would churn the same thought for hours, as a cow chews its cud. It did dawn on me that I had a sort of vacuum in my head for worries. I would worry about something, and whenever that issue was resolved a new one would pop into the worry spot.

When my mind wandered, there was a constant buzz of music in my head, which often consisted of phrases from a few choice tunes. I had no power to select them. There was "Wasting away again in Margaritaville," "I'm leaving on a jet plane," and "Silver threads and golden needles cannot mend this heart of mine, and I dare not drown my sorrow in the warm glow of your wine." Over and over again, the jukebox in my brain played the Everly Brothers singing "Wake up little Susie" and Jon Bon Jovi belting out, "Shot through the heart, and you're to blame; you give love a BAD name!"

In rare moments, the force of nature would assert itself upon my busy brain. Poetry had intimidated me in school, but in the vineyard I would be caught by words from a Dylan Thomas poem I had learned in high school: "The force that through the green fuse drives the flower/Drives my green age." I felt the same power in my veins, even though if I thought about it I was pretty sure that Thomas's poem had something to do with his penis.

I would stand in the vineyard after work, closely examining the vines' leaves for signs of fungus, and stroking the vines' tendrils. They would curl around my finger, responding to my touch. I thought that nature didn't care that I was a woman. In the evening air, I watched the moon rise and wondered if my emotions were affected by its gravity, or was it the tide in the nearby sea that sent waves of pleasure or grief over me?

Later, when I fell asleep, my work held sway over me as my mind kept going through the motions of pruning, tying, thinning, or picking. The muscles in my hand twitched until my thoughts became dreams, a seamless transition into the quiet void of deep rest.

It was one thing to be in tune with Mother Earth and another to be an Earth Mother. I had no old granny, no spinster auntie who would take care of my children when I was at work. When Anne was two, after I became pregnant with Zander, I put together a patchwork of baby-sitters that included sending Anne to Judy, a woman who took care of other children along with her own for the morning. After picking her up at two o'clock, I fetched a high school student who would spend the afternoon with her.

One day, when I went to get Anne, I got into a conversation with Judy. I was usually in a rush when I picked up Anne, but this time Judy was telling me a funny story about her son Brian, who was a little older than Anne, and I didn't want to interrupt her. It felt good to hang out a bit and have some girl talk. By the time I connected with the other baby-sitter, I was twenty minutes behind schedule. Until I was about to cross the railroad tracks that intersected our dirt farm road, I completely forgot that a train ran on those tracks every day at 2:20. I also didn't realize—although I had heard of the Doppler effect—that the sound of a train's whistle *follows* the train. Anyone in front of the train can't hear it very well.

There was deep mud on both sides of the tracks, and I was driving our delivery van, which had poor traction. I couldn't stop when I saw the train barreling down on us, so I gunned the engine. If that engine hadn't been a V-8, I'm sure that the train would have smashed baby Anne, the baby-sitter, and me to smithereens. There weren't more than six inches between the rear of the van and the locomotive. My heart almost failed me. Even now, I can't think of it without quaking.

Two months after Zander was born, I hired a gentle and loving woman to come to my house every day to care for him and Anne. She stayed for a year, but left to resolve a family problem of her own, right in the middle of our harvest. After she told me she was leaving, I walked into the field and begged Nancy, one of the pickers, to become my baby-sitter instead of my field hand. When she agreed, I felt I had been blessed with the perfect mother's helper.

Nancy was in her late forties with grown children of her own. She loved to do art projects with my children, to take them to the beach, and to read to them. She didn't clean the house or cook, but she would keep things in order and liked to put the laundry out on the line so that it would billow in the sun. The only thing she would not do was baby-sit on weekends—they were reserved for her husband.

After Nancy had worked for us for two years, both children were in school for most of the day, so she came only for a short time in the afternoon. She finally agreed to take care of my children on occasional Saturdays. One wintry Saturday, Alex and I were invited to have lunch with some friends, a psychologist and his wife, William and Barbara, who had two girls, Jess and Jen, the same ages as Anne and Zander. The older girls went to a movie at the library while Nancy minded the three-year-olds.

When Alex and I returned from lunch with our friends there was no sign of Nancy or of the children. The phone rang, and I picked it up. It was the librarian. "The movie ended half an hour ago," she said. "Who is coming to pick up your girls?"

Alex went flying down the driveway to get them, while William took off in his car to look for Nancy. Barbara and I stayed in the house in case they returned. Twenty minutes later William came back with Zander, Jen, and Nancy, who was completely inebriated. He described how he had found them while he was still on our farm road, going along the tracks about half a mile from our house. Nancy was carrying Jen, who was sobbing. Nancy had fallen on Jen in the snow, and the child was bleeding from a gash in her head. Nancy had gotten completely lost trying to find our driveway, and had gotten her car stuck on a mile marker in the middle of the field across the tracks. Little Zander was leading them home through the snow.

While Nancy was in the bathroom splashing cold water on her face, William told us that he had found about ten empty pint whiskey bottles in the back of her car. She was an alcoholic. It ripped me apart to let her go, but we had to. My children suddenly had the sad experience of losing someone they loved very much. They couldn't understand what was wrong with Nancy, although they knew she wasn't behaving properly. Just so they'd know that she was still a good person, and so she'd know we still loved her, I took Anne and Zander trick-or-treating at her house when Halloween came. It was the best way I could think of to visit her without making a big deal of it. It was weird to go there the way we did, with the kids dressed in their Halloween costumes as if they were fictional creatures. But how do you explain to a child that people are not always what they seem to be? We weren't what we seemed to be, either.

Our pioneering family was growing up, and while our closeness was no illusion, it was starting to chafe. Now we were not only growing grapes and making wine, but we also had to sell the wine, both wholesale and retail. While Alex anchored the office and the winery, I was the chief worker bee. I was everywhere that anything needed to be done. And as if there wasn't enough work to be done for our wine enterprise, tending children and animals besides, I insisted on adding a new wrinkle to the equation—I wanted our kids to go to school in the Hamptons.

When I was a high school student in the sixties I met a couple who were raising their child according to a philosophy that was popular at the time. The idea was to let children explore their world unfettered by direction or expectations. In this way, their natural curiosity would lead them to doing things that interested them personally. They would achieve at their own speed and be successful because they would be self-motivated. The couple's child was bright and inquisitive—just the way I hoped my children would be.

When I studied for a master's degree in teaching I chose to practice-teach in an "open classroom"—a setup that combined age groups and allowed the children to choose their own activities using a wealth of appropriate materials. This was exciting to me because, as a child who learned to read early, I myself had been bored in school. I didn't want my children to spend their day sitting at desks, filling out workbooks. When Anne reached kindergarten age I visited the local school, and that's what the students there were doing.

I researched the situation and learned that the Hampton Day School, thirty-five miles away in Bridgehampton, used the "experiential learning" philosophy that I wanted to follow. Although it was a crazy task to patch together transportation for the kids, Alex agreed with me that Anne, and then both kids, should go there.

Part of what was going on with this schooling fantasy of mine was that I was still deeply interested in education, and I wanted to be involved in a school where I could have some input. Another part was that I was looking for friends with whom I might have more in common than my old farming neighbors.

With so much going on there were few quiet times for Alex and me to be alone together, without spending the time talking about work. There was an undercurrent of anxiety because we really couldn't afford to do what we were doing. It was not unusual for visitors who came to the vineyard to look around the place, see the wine, the kids, the animals, and the handsome couple (us), and say to me, "I wish I had your life." I embraced this life, but there were moments that I wished I had what they thought I had—the pure enjoyment of it—too. I was just too tired to appreciate it all the time. Still, I was certain that by keeping up the good work, the pleasure that I genuinely felt more days than not would increase. It had to get easier.

❧

The baby-sitting problem seemed insurmountable, so Alex and I decided to forgo having an after-school sitter. Anne and Zander didn't get home from school until late in the afternoon anyway. We agreed that spending a little more time with our kids was part of the pioneering plan in the first place. Being in the house at the end of the day gave me more time to do household chores, and life did fall into a more comfortable rhythm.

Day after day, I woke up early, got Anne and Zander off to school, and went out the door to work. When the weather was cold, my work clothes—long underwear, extra socks, sweaters, wind pants, boots, parka, scarf, face warmer, hat, and mittens— created a cocoon of protection, a separate world that enabled me to travel safely through the harsh space of the whipping winds. At the end of the day, the house seemed all the more secure when

I pulled off all those clothes and wrapped myself, mothlike, in a comforter. Conversely, in summer, if it was hot, I would peel the sweaty clothes from my scorched and dusty body to stand in the shower under a stream of cold water and feel like Venus emerging from the waves. Never was sleep better than after a day outdoors.

Before that tumble into bed came the best part of the day—family time. We all had appetites for that evening meal. I cooked big dinners, letting the chopping and stirring bring revitalizing aromas to my senses. The food I cooked was simple, but since it was something I thought about much of the day, I wanted it to have some sensual quality. Calories were no issue, considering all the miles I walked up and down the vineyard rows. I liked to roast chicken with garlic and herbs stuffed under the skin. My osso buco was made with a flourish of orange peel. I braised fresh fennel until the inside of each morsel was creamy, while the buttery surface had browned to a nutty caramel. The kids liked spicy pork satay in a peanut sauce, and a spinach-ricotta quiche that I called "Popeye pie." A French intern who stayed with us one summer taught me his grandmother's recipe for shrimp sautéed in butter with *herbes de Provence,* flambéed with cognac. We had so much good fresh food from the farmstands and the fisheries that we ate different things in season all year round.

On Sunday mornings Alex and I made fresh bread together. Since I liked to get up early, I would proof the yeast and start a sponge of flour, yeast, and milk or water, so that by the time Alex added the rest of the ingredients and kneaded the bread, those yeast cells would be happily multiplying. Taking the ingredients that I had measured out for him, Alex approached the kitchen counter like Julia Child, towering over the bread bowl and using his enormous hands to work the dough with practiced abandon.

Desserts were strictly my territory. I always made something sweet for the end of the meal. Dessert was often pie, and there was a reason for that. Because of a bet that I lost one Saturday in June, I owed Alex fifty-six pies.

On that pristine summer day, we were invited to attend the wedding of one of our part-time employees. I read the wedding invitation and told Alex that we had to be at the Congregational Church in Orient, seventeen miles away, by the time the ceremony started at ten o'clock that morning. He tended to be a slow starter, while I hated to be late. At 9:15, I sat in the living room, fully dressed, my internal motor revving. Alex hadn't even taken a shower. He was in his study. "We are going to miss the wedding ceremony," I told him. "Get dressed."

The more I bugged Alex to get ready, the slower he moved. Finally, we were in the car going seventy miles an hour toward Orient. It was 9:50. I bitched some more. Alex turned to me with a challenge. He said that for every minute we were early to this wedding, I owed him a pie.

We got to Orient at 10:13. The organist was warming up for the wedding, which, she said, was scheduled for 11:00. When the bride walked down the aisle at 11:08, Alex won fifty-six pies. We laughed about my pie debt for years. I loved making pies, but there was one thing that wasn't funny. I didn't like being wrong.

At harvest time, I had a chance to bake lots of pies, and cakes, too. I got up early to bake for the picking crew, and after the last picking boxes of the day had been filled and before the sun went down, I would go to my kitchen again. I would make an enormous harvest meal for Alex and myself and any of the crew who were staying to finish the crush or the pressing. It was important to me to reward anyone who helped with the harvest, so that we could share a happy experience at a time when the work was especially hard.

Whether we made red or white wine, the fruit had to be stemmed and crushed as soon as it was picked. Red wine wouldn't be pressed until the fermentation was complete, because the skins needed to be in contact with the alcohol that formed during fermentation. This is how both color and flavor emerge in red wine. White wine, on the other hand, was fermented without skin contact, so it had to be pressed right away. The elements in grape skins, the phenols, that give red wine complexity are undesirable in white wine, as they can make it taste raw or bitter. A day picking white grapes meant a night of pressing, too.

Each press cycle could take several hours to run, since we wanted to use low pressure to avoid bruising the grape seeds in the press. Some presses extract over 200 gallons of wine for each ton of grapes, but we set our press to extract only about 160 gallons per ton. There were usually five or six of us who would work into the night. My cooking might be interrupted by a call to test the fruit in the lab or to prepare the yeast food and pectic enzyme, which was added at the crusher to help get the yeast going and to make the juice flow more smoothly. I'd stand on the pressing dock and ask Alex when he thought we could eat. "This press cycle will be done in twenty minutes," he might say, or, "We have to take the pomace [the dried mass of pressed grapes] into the field before we sit down."

I knew even before I asked that it was impossible to pin down the dinner hour. Even if it could be predicted, a trellis nail would jam the pump, a hose would split open, or the three-phase converter that powered the press would blow. I'd bring out a plate of cheese and crackers and wait to see what happened next. I longed for the moment when everyone would come tumbling into my warm kitchen, filling it with the aroma of denim drenched in grape juice. Then, there would be stories and laughter as the tension of the day disappeared.

Although the cook in me got upset when we couldn't eat as soon as the food was ready, the winemaker in me loved the sound of the press as it rotated with a steady whir. The press had a big "bladder" inside of it that inflated, pushing the berries together into a thick mass, the pomace cake. Periodically, when the pressure built up to a certain level, the rotations would stop, and the whirring sound would be punctuated by an alarming thump as the press retracted and the chains inside the press broke up the pomace cake. Spotlights on the press would make its stainless-steel basket shine. Through small slits in the cylinder, green juice or magenta must would pour in intermittent showers into the pan beneath it. Down in the winery, another pump moved the bright juice into a stainless-steel tank, where it would settle and continue its secret evolution into wine. It felt like a movie set—like a place where some fantasy was being played out. I pictured James Bond, trapped inside the press, stopping it from inside just before it turned him into juice.

In the evenings Anne and Zander kept me company in the kitchen while I cooked supper. Alex was usually in his study until the food was ready, working on bills, label designs, and marketing plans, or catching up on a little poetry reading. Anne would sit at the kitchen table, drawing, while Zander zoomed around underfoot. When he was eight months old, he had full mastery of his round walker. His first words after "Mama" and "Dada" were "own self." He was always fiercely independent. At two he became inseparable from his red plastic motorscooter, and at three he insisted on bringing his three-wheeled low-rider into the house. The cats sat on the old beams of the kitchen to watch him while he was in motion, but he never collided with anything. At six he took up soccer and kept that ball going inside and out, taking it to bed at the end of the day. At ten he decided tennis was his game, and if it was warm enough, he played against the winery wall. If it rained, the refrigerator became his backboard. If this all sounds tumultuous, it was, but it was also a happy sort of activity, a kind of ballet in the kitchen that went on for years without mishap or collision.

The only casualty of Zander's activity was the thirteen-inch TV that I had bought just before getting married. When he was two, Zander was fascinated by the earphone plug on the TV, and one day he tried putting the plug in too hard, and the thing fell over and broke. It didn't shatter; it just didn't work anymore. Alex and I decided not to replace it. It seemed more peaceful to read to the kids than to watch TV after dinner. Alex would go back to his study, while I took the kids up to the loft they shared upstairs. I would tuck them in and read or tell them stories about characters I invented, Blondie Girl and Elf. After the stories, I would sing to them. Our favorites were "The Sloop *John B.*" and "The Red River Valley." I also sang them my father's lullaby, something he had learned when he drove an ambulance for the British army during World War II and always sang to me when I was little.

> They say there's a troopship just leaving Bombay
> Bound for Old Blighty's shore,
> Heavily laden with time-expired men,
> Bound for the land they adore . . .
> They're all blinkin' barmy—
> Good-bye to the Army—
> And cheer up my lads,
> Bless them all.

I tried to sing that song to my children with all the love I had heard in my father's voice, and his relief at being safely home after the war. Every night the children begged for "one more story" or "one more song" but I soon heard them breathing slowly, softly, and I crept away.

10

Taking Care of Business

WHEN ALEX AND I GOT MARRIED WE AGREED THAT WE WOULD share in all our domestic tasks. As it turned out, there was more of a division of labor than I had expected. It made sense at the time. I would much rather be in the field, or cooking, or minding the kids than dealing with accounts receivable.

Alex spent long hours in his office making and implementing our business plans. He had to deal with the bank, order supplies, contact distributors, create promotions, fill out government forms, do the payroll, and take care of whatever else the business entailed. Besides that, for several years he wrote a weekly column for *The Hamptons Newspaper Magazine*. There were times when I wished he would take on more of the household chores, but on the other hand I saw how hard he worked on all the things I didn't want to do. When I thought about it—and I did, sometimes—I realized that a good feminist would not approve of the way I allowed the traditional male-female division of labor to persist in our marriage. But I could easily rationalize it. On the prairie in the 1870s Ma deferred to Pa, but they shared their troubles and their triumphs without bickering about who was working harder. They both worked hard, and so did we.

The only hint I had of Alex's mood came from the music he

played in his study. John Prine or Leonard Cohen revealed seriousness; Bob Dylan was for relaxation; Schubert's *Trout* Quintet signaled elation and a creative moment. With the kids and the pets and employees and equipment purchases, Alex had to be creative in funding this venture of ours.

Virgil and Columella didn't talk about money; they talked about taming nature. Money, I thought, is the antithesis of nature. Nature blossoms forth, while money runs away. Wherever it ran, it was up to Alex to chase after it. Our farm was mortgaged to the hilt.

As interest rates rose to twenty percent after the gas crisis of the mid-seventies, Alex had a great idea. He would offer Tom Polsky, the speculator who held our mortgage, an early payoff if he would discount the total amount we owed him. Tom lived just down the road from us, so we went over to his ranch house to negotiate. His wife, Hilda, offered us drinks and led us into their living room while she went to get Tom. As Alex, Tom, and I perched on the edges of their living room set, Hilda sat at an upright organ that had been purchased with some of the money they had made from selling us our farm. She began to play while Alex explained to Tom that he could take our cash and immediately make more money reinvesting it at twenty percent. That would be much better for him than getting the seven percent we owed him over time.

As Alex talked, the sound of the organ took me back in my mind to about 1954. I could see my mother ironing, while on the radio, chords of organ music built to a crescendo as the soap-opera stars professed their undying love for each other. The music changed to a minor key, and I knew their love was in danger. My mother never really followed those stories, and I was too young to understand them, but the sound of organ music still made me think of dire risk and changing fortunes.

When I tuned back in to Hilda and Tom's living room, Hilda

was playing "Easter Parade," while Alex's convincing voice rose over the music, obliterating any of Tom's objections to the scheme. With a check and a handshake, we paid off our mortgage. Almost immediately, interest rates dropped, and we took advantage of the vineyard's increased value to go to the bank and mortgage it again.

For several years we lived on the ten thousand dollars a year that Alex's grandmother gave him, plus Christmas and birthday presents from our parents. If they sent us something from a national retailer like L.L. Bean, I'd send the gift back for a cash refund. I bought clothes at the hospital thrift shop, or sewed them myself. If I wanted to give someone a gift, I would make it myself. I made rag dolls and stuffed animals for my nieces and my children; I grew flowers and herbs to sell for a little cash; I made jams and cakes to give at Christmas. We didn't buy books or subscribe to magazines or newspapers, so we were fortunate to have a wonderful public library in Cutchogue.

I didn't like to be reminded of how tenuous our situation was. The only thing I really wanted money for besides good food was the kids' education—which included the horse. Some of the time, Alex's parents footed the tuition bill. We bartered wine for lobsters and venison, and our neighbors continued to be generous with fish, fruits, and vegetables. Somehow we were always able to pay for essentials.

By 1978, our wines were selling well, and we had an opportunity to buy our neighbor's farm to the north of the railroad tracks. Wesley and Rose Simchick had been farming there for fifty years, but now they were old, and their sons wanted to leave farming. Their property had eighteen acres of good, flat land with frontage on the highway. They would sell us their large potato barn and keep their house, and they were willing to hold a mortgage. This would give us more storage space, plus an important retail outlet for our wines on the highway. We knew it would be hard to attract customers down our long dirt farm road when we had more wine to sell, so the purchase of the Simchick farm was essential for our business. Alex went over to talk to Wesley. Wes, as he was called, got red in the face, embarrassed to be talking about money. "What do you want for your farm?" Alex asked. Wes named a figure, and Alex agreed. That was the extent of the negotiation—there was no point playing games with modest, honest people.

We persuaded our partner, Bill Chapin, to approach his aunt Dorothy about this investment. She got a tax advantage by buying part of the property, five acres with the barn, and renting it to us. We retained the right to buy it from her at any time. We kept the mortgage on the farmland. Alex's mother helped us out by taking on a note backed by shares of the vineyard.

With Wes and Rose's farm, we got much more than a new place to plant Chardonnay. Wes and Rose became our friends. Rose taught me to make her special rhubarb pie and her anise chocolate roll. Wes could see that we were too busy to keep up the area around his old barn, so he mowed it himself every week, and sprayed our fruit trees gratis. If we needed to borrow a tool, he had a garage full of every new or obsolete handtool from more than fifty years back that we could choose from. If we needed a part welded, he would do that, too.

When he wasn't working around the yard Wes and his old farming buddies would go out to the bay on the motorboat that he kept parked on a trailer in his driveway. I would find them on a drizzly day, sitting in aluminum chairs outside his garage spinning yarns while they smoked the bluefish they had caught. Meanwhile, Zeus would be rolling in the fish guts that Wes threw into his vegetable garden as fertilizer.

Wes loved his vegetable patch. He always planted gladiolas

there for Rose, who treasured them. They had had a child who had died in infancy long ago, and I felt that those flowers were a memorial for their child. The row of spiky, colorful blossoms had a certain poignancy for me because I remembered that when my brother was born, when I was four, my father had bought my mother gladiolas. I can picture him clutching the bundle of emerging yellow blossoms as he took the car keys off the table on his way out the door to bring my mother and brother home from the hospital. I could see the look of relief on his face—now, after having two daughters, he had a son.

Wes had one of his own sons in mind when he planted an eighth of an acre in lima beans every year. For two weeks at the end of the summer, he would sit in one of those aluminum chairs by the garage, shelling the beans. He always urged me to take some, but I declined. "Lima beans are the one vegetable I don't like," I told him. "They're too furry."

"I don't like them, either," Wes said. "I grow them for my son."

The next time I saw his son, I told him that I thought it was wonderful of his father to grow and shell all those lima beans for him, considering that he didn't like them himself. His son looked surprised and replied, "I hate lima beans. I just take them to please Dad."

A few years after we bought their land Rose found out that she had cancer. She went through a course of chemotherapy that nearly killed her, but after it she lived for several more years. When the cancer returned with a vengeance, she succumbed in a matter of days. At the funeral home, where I went to pay my respects, I looked at her rouged and plastic-filled body lying in the casket. Whatever was left of her bore no resemblance to the warm and loving woman I had known. How awful it is, I thought, when the living force is gone.

After Rose died, Wes kept on mowing our grass and welding

our tools, smoking his fish and shelling his beans. He kept the gladiolas growing, too. When I stopped to visit him, he would tell me about his aches and pains in a matter-of-fact way, always ending the conversation by saying, "It could be worse, like riding in a hearse."

🜲

Preserving the life of our vineyard and helping it to grow became all the more compelling to me. It was more than a matter of planting and tending the vines. The business had to be kept alive, too. Once we had wine to sell, we had to figure out how to market it. In 1988 Alex came up with a way to sell wine before its time through wine "futures," so that we could afford to buy French oak barrels. We offered to sell our red wines to our customers a year before they were bottled, at a reduced price, which would give us time to age them properly. We represented our wine as being French in style, and we felt that the quality of our wine should live up to that comparison. Those American oak whiskey barrels we had used for our first vintage just weren't going to do it.

Appearances are deceiving when it comes to oak barrels. Put an American white oak barrel next to a French white oak barrel, and they look the same. Zoom into a cross-section of the wood with an electron microscope, and the difference is obvious. The American oak, *Quercus alba,* has a much denser structure. It can be mechanically sawed into staves that are kiln-dried, using at least twenty percent more of the tree than French oak *(Quercus sessilis* or *Quercus robur),* which has to be hand-split. All the harsh tannins and strident woody flavors of the American oak barrel are quickly leached into any wine that's stored in it. French oak is more mellow, and much more expensive.

Maybe the American oak was more of a match for how we really were ourselves—up front, full of bravado. But that's not how we wanted to be. If we thought we could make a success of the vineyard in a hurry, we were learning that the wine couldn't be rushed. Depth and nuance take time and maturity. Oak staves need to be cured outdoors in the cooper's yard so that funguses—the same funguses that make Roquefort cheese—can change the nature of the wood itself. Sun and rain alter the stridency of the wood before it is shaped for the barrel. Trial by fire—the fire that the cooper uses to bend the staves—caramelizes the wood to impart a sweetness to the wine. It takes two to three years to prepare wood for a French wine barrel.

Once the wine is in the barrel, then the maturation process begins for the wine itself. The rough inner surface of the wood is like the coast of Maine, with many coves and inlets. Wine molecules bump into each other along that coast, forming chains that change the harsh, astringent tannins into round, soft, interesting aromas and flavors.

When I think about it, Alex and I were going through a similar maturation process—but the effects of nature and circumstance were not all mellowing for us. They made us vulnerable, just as an oak barrel—any kind of oak barrel—leaves the wine susceptible to spoilage if air or the wrong microbes get into it.

Fortunately, our wine-futures program was a great success. Before Christmas every year, Alex would design an elaborate mailing for our customers. They would have to come to the winery the following October to pick up their wines, and usually they would spend more money when they came. It was a brilliant way to build a loyal customer base of wine lovers who were guaranteed to get our best.

Creating the mailings was an enormous undertaking. For several years in a row, Alex invented an entirely new offering. One year he created a full-sized poster with photos of people and places that were meaningful to our wine production or to life on the North Fork. Another year he wrote a synopsis for a vineyard opera and drew the divas himself. The year after that he designed a calendar with descriptions of how the vines grow and the wines develop in every month. We had folded eight thousand of them before a visitor looked at them and asked, "How come on this calendar November comes before October?"

Anything that had to do with design became a project that Alex would throw himself into. Before our first wines were released in 1977, in order to create a wine label, he selected the wine labels he most admired from all over the world. Soaking them off their bottles, he pasted them onto the wall of the tractor shed so that he could study them over time, at a distance. We agreed that the label should be both traditional and New World, but how would we achieve that?

First, we hired a graphic designer who produced several samples, all of which we rejected. These labels were entirely text, and we wanted an image as well. We talked about the idea of using a drawing of a burgeoning grape shoot, symbolic of the growing wine region's potential. Neither of us had the skill to do this drawing, so we went for help to a local landscape architect, Rob White, who was also an artist. As Alex sketched his idea for the vine, he drew a line under the shoot and said, "All the printed information goes under here." Then, looking at the line, his face lit up. "This can be the trellis wire," he said. "The shoot can hold on to it with a tendril."

I went into the vineyard and picked a grape leaf for Rob to draw. We put it together with type from an old foundry face, and Long Island wine's new image was born.

The printer we used housed his business in an aging industrial building, strangely located in a wooded area where no one would ever expect to find industry. I had Anne, age three, and baby Zander in the car with me one day when I went to pick up our new labels. The printer's name was Lionel, so when we arrived there I said to Anne, "Let's go see Lionel." She started to wail at the top of her lungs, and refused to get out of her carseat. It took me several minutes to figure out that she thought I was taking her to see a lion. We had visited the zoo a few weeks before, and she had been frightened by a real, roaring lion.

To decorate the Simchick barn that we were renovating as a visitor center, Alex went to salvage yards where he found wrought-iron gates and stained-glass windows. The place, known to us as "the New Barn," looked like a potato barn on the outside, but when visitors entered, they saw the full-sized stained-glass depiction of Jean-François Millet's *The Sower* and the wrought-iron balcony that was covered with grape leaves and tendrils. They would often stand for a moment to get used to the dim light, sigh, and say, "This is a temple of wine." With its elegant Victorian table—inherited from Great-Aunt Helen—and its oak paneling that our partner, Bill, had salvaged during his renovation of the Roby Tool Company in Rochester, the space captured our vision that besides being a delicious beverage, wine had aesthetic, spiritual, and monetary value.

※

In the early years of the vineyard, Alex was tireless in trying to present our vineyard and our wines in unique ways. Whatever he did, he did with total immersion.

There is a passage in Virgil's *Georgics* that was a favorite of Alex's and symbolic of this immersion. He kept a leatherbound copy of Dryden's *Virgil* opened on the table in the tasting room to the verses:

> *Great father Bacchus to my song repair;*
> *For clustering grapes are thy peculiar care:*
> *For thee large bunches lade the bending vine;*
> *And the last blessings of the year are thine.*

> *To thee his joys the jolly autumn owes,*
> *When the fermenting juice the vat o'reflows.*
> *Come strip with me, my God, come drench all o're*
> *Thy limbs in must of wine, and drink at ev'ry pore.*

One night, when the harvest moon was full, and the large outdoor tank—the one that didn't fit into the building—had just been charged with red grapes that filled the vat to about four feet in depth, Alex took my hand and said, "Let's do it."

We tossed our clothes on the slope that hid the tank from the house and climbed naked through the manhole into the red, red must. The air was cold and the seeds and stems scratched my flesh, but we danced and danced on the seething juice until the dance was over and we returned, subdued, to bed.

11

The World at Our Door

ALEX AND I WERE SITTING IN THE KITCHEN EATING LUNCH one day when our neighbor at the end of the drive, old Mrs. Zuhoski, called. "There's a man who's stuck in the farm field, says he's looking for you." Alex jumped on the tractor and roared down the road to tow yet another miffed and muddy news reporter out of the field.

Barely weeks after we planted our first vines, Barbara Delatiner of the *New York Times* came to report on our pioneering venture, and while she didn't get stuck at the end of the road, about twenty percent of the professionally and amateurishly curious people who came to see what we were up to did. After a while I started to tell them, "No matter how bad it looks, stay on the driveway. The field looks smooth, but it's quicksand."

In 1975 Berton Roueché of *The New Yorker* somehow got wind of Alex's interest in Virgil, and came out to report on how Virgil could teach a couple of kids how to grow grapes. He sat at our picnic table under the maple tree, interrogating Alex and eating a quiche that I had made while I tried to keep baby Anne busy with her favorite toy, a plastic lab bottle full of her barrettes. Alex came up with one bon mot after another, and

Roueché, taking notes longhand, couldn't keep up with him. Every time Alex said something interesting, Roueché would scribble madly and then say, "Could you repeat that?"

When it came to entertaining the news media, our motto was "Bear and forbear." We needed them. We had no money for advertising, and while word of mouth about our wines was effective, it was slow. We were in uncharted territory economically. Just as the homesteaders on the frontier had to try to predict how many dried beans, pieces of salt pork, bolts of calico, and Indian trade blankets they would need to take to the prairie, Alex had to calculate our financial needs. What would it cost us to expand our plantings, start the winery, market our wine, and feed us something more than the eight-foot-high pile of potatoes that were stored in our neighbor's barn? Since everything we had was already invested in the land, he had to keep the bank and our partner, Bill, happy. He and I both had to make believers out of anyone and everyone who could possibly help us.

In November 1977 we got help from the media in an unexpected way. I was dressing baby Zander when I glanced out the window and saw a yellow New York City taxicab coming down our road. Quickly bundling up Zander, I went outside to find Alex gesticulating toward the vines to a short Vietnamese man who was unloading boxes of camera equipment onto the pressing dock by the winery. "This is Ted Thai," Alex said. "*Time* magazine sent him here to take some pictures. Can you show him the wine cellar?"

The taxi waited, meter running, for the rest of the day while Thai set up his shots. After Thai saw the cellar, he scowled and said, "Yellow. I need yellow."

There was nothing yellow in the winery. I didn't have any yellow clothes and neither did Alex. His brother, Charlie, had a yellow raincoat. Thai said that was fine. He set up his tripod, reflectors, lights, and meters in the cellar. Now, he asked, could Charlie make some water splash? Wearing the yellow coat, Charlie stood in the cellar for over an hour, pretending to wash a barrel with a pressure washer that sent a fine mist all over everything. "But this has nothing to do with winemaking," I complained. Thai didn't care. When he finished with that shot, he said, "I need faces with character. You are too young. Go to town and find me faces with character."

That was the limit. Alex removed himself. I told Thai that he would have to settle for my rapidly aging profile. I had Zander on my back, and he was wearing a rainbow-colored sweater. In the lab, Thai set up a shot of us pretending to run a test that creates a neon pink liquid when it's done. If I couldn't give him a face with character, at least I could give him color.

In halting English, Thai told me that he had no interest in taking what he considered to be snapshots. He had done enough of that in Vietnam. *Time* magazine had evacuated him from his home there when the American troops left Saigon. "I was on a gunboat on the Mekong River, taking pictures for *Time*," Thai explained. "The soldiers on the bank started firing at us, and the boat swerved. I had to choose—grab the hot gun mount next to me, or fall in the river." Thai extended his hands and turned the palms up so that I could see the scars. "All this flesh—burned," Thai said solemnly.

A few weeks later, *Time*'s article appeared. Under the title "Shaking California's Throne," there was a view of Zander in his rainbow outfit, looking wide-eyed at the camera from my back while I intently pipetted the lab sample. The caption was "Impressive hints of the future." On another page, Charlie in his yellow coat lent a glow to the mist that looked like fine gold pouring over the barrel. The text gave our 1976 Sauvignon Blanc a "top rating among New York wines," and we were launched. Alex's father finally loaned us some money;

the bank upped the ante, and our partner, Bill, sweetened the pot.

The following year, Frank Prial of the *New York Times* wrote a glowing article about us in the *Times Magazine.* None of us paid much attention to his final paragraph, which read, "Because they are so young, so successful, so delighted with their life, there is an aura of purity and innocence about the Hargraves that is appealing but disconcerting to anyone who has dwelt too long in a world where nothing is what it seems to be. It's as if they live in another world, one that most of us will never know. A world where no one has ever failed."

As far as I was concerned, there was no room for failure. Every year, I rejoiced that nature gave us a fresh chance to try again, to correct our errors of the year before. The vineyard was still an experiment after all. But as time went on, I became frustrated that we always seemed to be behind schedule. Even when our wines got great reviews, they were expensive to make and to market. It bothered me that although the *Time* article had featured a photo of me, it was only because Alex had absented himself from the photo shoot. When other reporters came out they only wanted to speak to Alex. I might as well have been invisible. When I gave tours, inevitably someone would ask, "So, when did your husband decide to grow grapes?" I spent hours with visitors who wrote Alex—not me—thank-you letters without ever having met him. Although Alex always included me when he talked about our vineyard, I felt marginalized by the rest of the world.

I believed that if we planned a little better, we could get our work done ahead of time. That would save money on labor, and keep our equipment from getting beat up from having to deal with the tangle of vines and weeds that we always had by August. I liked Alex's creative work; in fact, I was in awe of it. But sometimes, when he was toiling in his study in the middle of the night, I wished he would quit being so creative. While he was designing labels and inventing opera scenes for a mailing, those weeds kept on growing out in the field.

One April I told Alex that, for once, I wanted to be ahead of the weeds. We had a couple of people helping us, but if we had just one more person to help in the field, we could get the vines tied early enough for the tractor to push the wood out of the vineyard; we could tighten the wires and start cultivating *before* the weeds were ankle-high. "Okay," he agreed. "See who you can hire."

The very next day, a young woman appeared at the farm looking for work. She was cheerful, smart, and said that she was an active athlete. Not only that; she said she could work for just over minimum wage—less than any of our other employees. I jumped for joy and took her to meet Alex. I thought I saw him wince, but I couldn't understand why. This girl was perfect! But Alex wasn't sure she was tall enough to do the work. We went into the vineyard to see if she could reach the top wire.

As the three of us walked into the vineyard with the tallest trellis, I took Alex aside and told him what the woman would work for. I was absolutely gleeful by then. The woman obligingly reached up to the top wire, which was at about six feet. She was no shorter than another woman who had worked for us for years, but Alex shook his head and told her that he didn't think she could do the job.

It was just a small thing—not hiring this woman—but the rage I felt at that moment seethed deep inside of me. I knew that instead of one cheap employee now, we would have to hire five or six in August to put up shoots and pull out weeds. Instead of getting the cultivation done early, with the tractor, it would have to be done by hand. And I would be the one doing it myself. I was completely exhausted, taking care of the kids, the animals, the cooking, and the housework; working in the field, the winery, and the store. All I wanted was a little help, and I didn't get it. I didn't want to have a fight with Alex about it, either. I was too tired to fight.

I think that Alex was terrified to take on another employee at any price. In the springtime it was hard to predict how the new wines would turn out or how well they would sell. The excitement of pioneering was giving way to the drudgery of trying to do the same thing over and over again; meanwhile the vast open spaces were closing in.

We lived almost as if we were inside a fortress. There was too much going on that we had to deal with. We were so close, literally and figuratively, that it was safer for us to get annoyed at each other than at anyone else. It is easy now to see the signs that our marriage was headed for trouble, despite the deep and abiding love we shared for each other, our children, and our vineyard. Some of the personality traits that made us interesting to each other at the beginning of our relationship became annoying differences as time went on. For the business, it was good that I was an early bird while he was a night owl. There was always someone alert enough to keep working. The other side of it was that one of us was cranky when the other one was full of beans and vice versa. I wanted to schedule everything, but Alex liked to see events play themselves out, which made me a nag and caused us to miss some opportunities. Most fundamentally, while initially I often wanted Alex to be the one who made choices for the family, after I gained a sense of my own strength and competence I wanted to take on more responsibilities—but then again, I didn't. Still, every time things got too uncomfortable, we were able to talk it out, to get comfortable again. I think it's fair to say that we both wanted our marriage to succeed more than anything.

The future of our marriage was inextricably tied to the success of the vineyard, too. For that, we needed to broaden our horizons beyond the North Fork—into the realm of Manhattan and the Hamptons. Imagine a prairie chicken entering a roomful of fancy fighting cocks and saying, "Hi, guys. Wanna play?" That's about what it felt like.

We did several things to equip ourselves to emerge from the country bumpkin stage to being in a position to offer our wine to a sophisticated market. In 1979 we spent our last five thousand dollars on a down payment for a Mercedes-Benz. We got the smallest model, the 240D, known as "the Berlin taxi" for its rattling and indestructible diesel engine. Five-year-old Anne went with Alex to buy it and picked out a dark blue one that Alex named "the Bluebird" after his father's favorite song, "The Bluebird of Happiness."

I was invited to join a prestigious professional women's food and wine society in Manhattan, Les Dames d'Escoffier, and Alex was invited to join a board of community advisers for the Parrish Art Museum in Southampton. The Parrish asked us to donate our wine to one of its events, and Alex got the museum to agree to let us use a gorgeous painting, *The Bayberry Bush,* by William Merritt Chase, on our first Collector's Series label. The following year, when Anne became a student at the Hampton Day School, Alex joined the board of directors and was elected treasurer of the school.

We weren't exactly out of our element in either Manhattan or the Hamptons. After all, we had both gone to Harvard. I could trace my heritage back to the *Mayflower,* and Alex's grandfather had been a captain of industry. If we had continued on the path we had been raised for, we wouldn't have been any different from the people who were serving our wines at their glamorous parties. Nonetheless, after a few years of carrying babies on my back and working outdoors in every kind of weather, I felt more relaxed eating Good Humors with my kids in the yard than dining on lobster in an oceanside mansion, where many of the guests were making use of the little bowls of cocaine that decorated the mantel. We were different from the farming neighbors who included us so graciously in their lives, and we were different from those who lived the high life in the Hamptons, too.

One Halloween Alex and I were invited to a party given by a New York couple who were famous for both their wine collection and their parties. The party was at their home in the Hamptons. The invitation had said, "Attire: witches and werewolves." I looked around for something black to wear, smudged my face with charcoal, and stuck some feathers that I found in the yard in my hair. Anne, who was about ten years old at the time, looked at me and had a fit. "That's not what they mean by 'witch,' " she said. "They mean slinky with red lips."

I knew she was right, but I wasn't slinky. I scrubbed my face and put on red lipstick, though. At least I didn't look like a freak, Halloween or no Halloween. It was easier for Alex to dress as a ghoul for the party. He wore his grandfather's tails and a noose around his neck.

The party took place in the basement of our hosts' pre-Revolutionary house. There were nylon cobwebs and fake spiders on the old stone walls, and dramatic music to make the place spooky. Sure enough, our hostess looked fabulous in a tight dress with deep cleavage. She served roasted corn with plenty of garlic—in case any of the werewolves were real.

I don't know if it was because Alex was suave or because he was a man with an advanced knowledge of wine, but there was one great party that he was invited to and I wasn't. His hosts were a couple, Barry and Lola, who were growing wine grapes as a hobby on their estate. Lola was a movie star and Barry was a broker, and their house was a castle with a moat. We had met them at some wine-related event and liked them very much. One of Barry's associates, a great wine aficionado, was having a birthday party at their castle and had invited the chef of a famous restaurant to fly in by helicopter for a wine dinner. Some well-aged wines from famous French vineyards were also brought out for the dinner. It devastated me that I wasn't included, although I understood that Alex was intended to be an escort for the well-known restaurant reviewer Gael Greene. He was seated next to her, and told me that when the other guests let their praise for the wine exceed its merit, Gael turned to him and said, "If you ask me, this wine smells like a horse and tastes like a saddle."

The next time Barry and Lola invited Alex to a party they included me. Even after they pulled out their vines in exasperation with the birds, deer, 'coons, and wasps that ravaged them, Lola and Barry were still very interested in the wine world. I thought I held my own well enough against many of her glamorous guests, but I was no match for one of them—a French model whose husband had a famous line of jewelry.

We had met this model, Mignonne, a few times at various parties. She took her gigantic black personal trainer with her everywhere, and could usually be seen in as little clothing as she could possibly get away with wearing. She had lived in Hong Kong, where she told me, it was customary to give gifts of gold to celebrate important occasions. At cocktails, she came up to me and said, "I see that you are not wearing a watch. I will give you mine. Then my husband will have to give me a thinner one." Before I could say anything, she had taken her solid gold watch off her wrist and put it on mine. I didn't know what to do. I didn't want to insult her by returning her gift. I remembered that my mother had made me return a rhinestone bracelet that my classmate had given me on the bus ride home from second grade. She had said that it was hideously garish, but I had thought it was the most beautiful thing I had ever seen. I was generations away from my Presbyterian missionary forebears, who had kept everything plain and simple—Why did I have to live with their legacy of austerity? I hadn't wanted to return the rhinestone bracelet, and I didn't want to return the watch now. But it was outrageous—Mignonne hardly knew me! Shaking my head, I muttered thanks and gave the watch back to her. She went off to fish in warmer waters.

In the winter of 1982 Alex and I opened our eyes to find ourselves in a fairy-tale bed with ironed linen sheets, being served fresh croissants and café au lait by a demure French maid who curtseyed as she left our tray and exited the room. We were in a private guestroom at Château Pichon-Lalande, a *grand cru classé* vineyard in Bordeaux. Looking at the damask draperies, the silver bibelots on the dresser, and the view of neighboring Château Latour, we asked each other, "How did we get so lucky?" This was what the wine world was all about. If the New York wine industry was petty and narrow-minded, the French welcomed us with open arms.

The fairy godmother who had invited us to her castle was its owner, May-Eliane de Lencquesaing. May had inherited the château in a family lottery that took place in 1978, and she was modernizing it with the dedication of a mother who is helping her child recover from a near-fatal illness, and with the organizational abilities of a commander in chief. Her husband, Hervé, was a retired general in the French army who had gotten his stripes in Vietnam. He and May had been stationed in Kansas in order to help the U.S. Army for a couple of years, and both of them spoke English.

Having lived in the United States, May understood how valuable the American market was for her wine. She had come to New York on a promotional tour in the winter of 1981, and a reporter for *Newsday,* the Long Island paper, had brought her to see our operation. As he described it, "The countess lifted her tulip-shaped glass and held it to the light filtering through the Tiffany stained-glass window rendering of Millet's *The Sower.* She swirled the deep red wine, tasted it and looked up at Louisa

Hargrave. With a puckish smile the countess said, 'It's against my interest, but this is a Bordeaux.' "

May-Eliane was not a countess, but she told me that if people confused her with the Comtesse Sophie de Lalande, who used to own the château in another century, she wasn't going to correct them. She knew how much Americans are impressed by royalty, and she certainly could carry herself royally. Maybe because she sensed in Alex and me the same determination to prove ourselves that she felt in her own situation, she took us under her wing and insisted that we visit her at her château the next winter. I think she was impressed that we had invested in stainless-steel tanks, something she was planning to do for her winery. At the time, many Bordeaux wineries were in dispute about whether these tanks were better than the old, open-topped wood fermenters, or the concrete vats they had "modernized" with in the fifties. May agreed with us that because they are more sanitary and allow the wine to ferment under controlled temperature, the stainless-steel tanks were an important improvement.

When we discovered that Pan American airlines would give us free tickets to Paris if we would do a wine tasting in the first-class cabin, we booked a flight. We didn't even have to do much of a tasting, since most of the passengers in that class were Arabs who didn't drink alcohol. The customs officials at the Paris airport didn't know that Alex and I could understand French; they argued with each other about whether or not they would permit us to bring the wine that was left over from the tasting into France, and if so, what they should charge us in duties. When they learned that our wines were from New York, not California, they lowered their lips in disdain, said, *"Eh, beh!"* and passed us through with our wine.

We drove with Hervé and May from Paris to Bordeaux, and by the time we had reached the château, we felt like family. May took us with her everywhere for several days, so that we could see all the improvements she was making. At an antique shop, she bought us two porcelain plates that were typical of the region. She invited us to lunch with Émile Peynaud, the man who consulted with Bordeaux's top châteaux. Peynaud told us that his concern was to make sure that each château that he consulted with would improve its wine quality but retain its traditional identity. When Peynaud tasted the wine we had brought along from Long Island, he agreed with May that it was truly like a Bordeaux.

May said that she was glad to have us stay with her because it gave her an excuse to visit her neighbors' wineries and see what they were really doing. As the hostess of American winemakers, she could come along and listen to how they answered our questions about equipment and technique. Since she had owned Pichon for only a few years, she had to learn everything about wine, and learn it immediately. She had a good winemaker, but she wanted to know for herself what it was all about.

On the desk in our bedroom at the château there was a book about Bordeaux wines that listed specific techniques used by each winery. We were going to visit Château Cos d'Estournel, and I noticed that the author of the book said that the château didn't filter its wines.

Filtration is a technique that has forever been instrumental in winemaking, whether using canvas sacks, asbestos sheets, or, in recent times, cellulose pads. Most wines have a certain amount of suspended particles that impair the brilliance and even the flavor of wine, and filtration is a useful way to clarify them. Modern winemakers have many filtration techniques at their disposal. A gross lees filter uses a coating of diatomaceous earth on fabric pads, under high pressure, to separate the lees (the gloppy fruit particles or yeast cells that fall to the bottom of the tank) from a significant amount of wine that would otherwise be lost when the lees are discarded. This kind of filter is used in white wines more than red. All wines may be filtered again at other stages, usually with a series of pads that allow a specific size of molecule to pass through. In sweet wines, or wines that have not been naturally stabilized by the malolactic fermentation, a cylindrical sterile filter can be used to ensure that no microbes at all will remain in the wine as it is bottled.

It has become fashionable to claim that high-priced wines are "unfiltered," mostly because filtration can strip a wine of its depth and complexity if it is used at the wrong time, or more stringently than necessary. This is especially true of fine red wines, since they rest in barrels for over a year, and the tannin of the barrels, plus gravity itself, effectively clarifies the wine. But saying a wine is "unfiltered" can be misleading, since there are other methods—like using a centrifuge or various fining agents—that can clarify wine but at the same time strip it more than filtration.

We used a filter when we bottled our wines, so I was especially interested to learn how the winemaker at Cos could avoid doing the same.

When we got to Cos, we entered through the back of the *chai*. Right there, the staff was putting the wine through a plate-and-frame filter, just like the one in our winery. We met the winemaker, and I asked him if he filtered his wines. "No, no," he replied. I pointed outside at the filter and asked, "What's that?" "It's a filter," he said. Then he explained to me that a wine is considered "filtered" only if it goes through a sterile filtration on the bottling line. Well, I was mighty glad to have that issue cleared up.

❧

The last night of our stay at Pichon, we started to discuss the possibility of a joint venture on Long Island. May had some money outside France that she wanted to invest, and she was impressed with the quality of our Long Island wines. It was the cook's night off, so we rummaged around in the old château kitchen for some macaroni to eat with cold sliced ham and some oysters that Hervé had bought in town. While sipping a bottle of '79 Pichon, we agreed to form an equal partnership that would be incorporated as Domaine de Lalande. They would own the land, and we would do the work.

At the end of our supper, May and I cleaned up together in the pantry. She filled the sink with suds and washed the plates and utensils, handing them to me to dry and put away. When she was done, she drained the sink, scrubbed it, and carefully wiped it spotlessly dry with a dishcloth. This small housewifely routine was symbolic of the way May operated. She could have left the cleaning for the maid, but she didn't. The way she wiped the sink impressed me as much as the way she wore her Chanel suit and kept her hair perfect. Here was a woman who was as thorough at dishwashing as she was in restoring her château. She was a military wife who was becoming a noblewoman, and she had no problem playing either part.

When we got back to Long Island, Alex got to work making a legal corporation with the de Lencquesaings. To raise the money that would permit us to go forward, we sold Suffolk County the development rights to our vineyard. That meant that no one would ever build a subdivision of little houses on our land, and it would remain a farm in perpetuity. Now we were more committed than ever. We bought our first automatic corking machine, and began giving tours and tastings at the winery.

❧

I was working on the bottling line the following summer, on a particularly hot and sticky day. The stereo speakers that had been an engagement present from Alex's parents back in 1968 were tied onto the rafters in the cellar, and they were blaring country music. It took four people to work the bottling line: one to open cases and hand bottles to the person who put them on the spigots of the filler, another to take them off the filler and cork them, and a fourth to case the filled bottles and stack them on pallets. I was putting a bottle into a case when I saw something white out of the corner of my eye, and looked up to the top of the cellar stairs to see a clean-shaven, pomaded young man dressed in a pressed white linen suit standing there, watching us. Stopping the line, I went up the stairs to greet him. I was wearing old khaki shorts and a T-shirt that was drenched in wine—it was eau de Chardonnay that day. The wine smells from the cellar didn't prevent me from noticing our visitor's own subtly complex cologne.

In perfect British English with a hint of a French accent the young man introduced himself. "I am Henri-Melchior, May de Lencquesaing's friend from Paris," he said, handing me a large calling card. I looked at the card, which read "Prince Henri-Melchior de Polignac." Inviting Henri-Melchior to join us for dinner after the bottling was finished, I sent him to find Alex in his study while I finished my work. At dinner Henri-Melchior explained that his family had recently sold its interest in Pommery Champagne, and was looking for another type of wine investment. May had told him about us, and he wanted to know if we could help him find land for a vineyard on Long Island, too.

Over the next several months, Alex worked on a plan that would suit Henri-Melchior and his family. Henri-Melchior's parents came to our farm to see it for themselves. His mother, Gladys, was an American-born heiress who had been active in the French Resistance against the Germans in World War II. Prince Guy, his father, was a cousin of Prince Rainier of Monaco—the one who married Grace Kelly. Prince Guy had grown up in an enormous palace on the Seine and would have been somewhere in line for the French throne if the French still had a monarchy.

I don't know if they had ever been in a house as humble as ours, but I wasn't worried about it. Food and wine are great equalizers. I made them a lunch of soup, salad, and pears baked in honey, nothing too pretentious, but I made the soup with local bay scallops in a broth of saffron and cream. I had grown the lettuce, and the pears were from Wickham's farm. We toasted each other with our '81 Chardonnay, one of the best we ever made, a wine with the energy and roundness in the mouth that a maker of Champagne would appreciate. Prince Guy stood at my stove, holding his glass of wine and stirring the soup with a wooden spoon. I saw him inhaling the aroma of the soup, and then sniffing the bouquet of the wine. When I invited him to sit down while I finished preparing lunch, he said, "No, thank you. I want to stir the soup. Nothing could make me happier."

As promising as our deals with the French were, both of them fell through. May had us vetted by a financial partner of hers from the Philippines and a New York City lawyer. We sat in a smoky conference room at a midtown New York hotel and, without the aroma of soup and wine to win over the vast skepticism of these grim men, the deal fell apart.

The de Polignacs found that they couldn't get their money out of France under the new socialist government, so they bought a *cru bourgeois* château in Bordeaux (Château Liversan) instead. We took our children to visit them in France in 1984. After the trip, Zander, who was six, insisted that the bald African man-servant who served us lunch at their residence in Paris was wearing only a lion skin. I guess it was a little overwhelming for a six-year-old.

Several years later, Alex and I went to Bordeaux again at the invitation of May de Lencquesaing to join a group of American winemakers on a tour that she had organized. We knew that Princess Gladys de Polignac had died, but we were looking forward to seeing Henri-Melchior and his father again. Just before we left for France, we heard that Henri-Melchior was dead, too.

As soon as we reached Bordeaux, we called Prince Guy to arrange to see him and give him our condolences. Château Liversan was a very quiet place this time; there was no bustle of construction and renovation as there had been before. Prince Guy emerged from the château to greet us, his face ashen and his shoulders slumped. "I am desolate," he said, tearfully. "Henri had flown his small plane to the château to visit me for the weekend, and I insisted that he take my secretary up for a ride over the vineyard. I had promised her this ride months before, and Henri had always said that he was too busy. This time, he decided that he could make a video of the château from the air while flying the secretary around, and kill two birds with one stone. He must have been looking through the video camera when his plane flew into the electrical cables. They died when the plane hit the ground and burst into flames."

Prince Guy led us into the vineyard and showed us where he had planted flowers to mark exactly where the bodies had been found. He asked us, over and over, if we had any idea whether Henri might have had a child—a bastard would be fine. He was desperate to leave his château to a son, and although he had daughters, they had been raised by other families during the war. He didn't think they would want to keep the château going. Prince Guy kept asking us for all kinds of advice. We left him bereft of all he cared about and unable to fill the hole in his heart.

❧

It was the summer of 1985. Alex came leaping down the winery stairs waving a piece of paper. "We won! Our Collectors Series

Chardonnay won the Prix d'Honneur at the International Food Show!" he announced.

This wasn't the first award we had won, but it was the biggest. The next year we were thrilled to hear that our '85 Chardonnay and '83 Cabernet Sauvignon had won gold medals at another major wine competition. That is, we were thrilled until one of the judges called Alex to report on what had happened during the judging. According to his report, any wine that had won a gold medal (as ours had) would be put forward for the Best in Show award. In the past very fruity, sweet wines had always won this particular contest. When our Cabernet, a very dry red wine, was tasted blind by a panel of judges, this judge told us that the panel agreed it deserved to win Best in Show. However, as soon as the identity of the wine and the fact that it was from Long Island were revealed, another judge on the panel had a fit and blackballed it. He didn't want a Long Island wine to win—and so it didn't.

Alex thanked the judge who called him. We took our medals, draped them around some empty bottles in the lab, placed them next to my jar of Pinot Noir in formaldehyde, and never entered another wine competition.

We found that we didn't need to win medals anyway, because over the years, our wines got great reviews. In 1984 even the eminent Robert Parker stuck his neck out to write in his *Wine Advocate,* "Both the 1982 and 1983 Fumé Blancs have a remarkable resemblance to the exceptional white wines of Graves of Domaine de Chevalier. . . . The 1981 is one of the most exciting domestic Cabernets I have tasted in the last year. . . . The expansive perfumed bouquet of cedar, spicy oak . . . suggested the style of one of Bordeaux's finest St.-Juliens."

All of the wonderful words that were written about our wines were gratifying. They meant that we were doing what we had set out to do. As time went on, however, I began to feel somehow inauthentic. Yes, the wines were delicious, but something was missing. I don't know if it was missing from the wines, or missing from myself, but no matter what anyone said—or what I said, for that matter—the wines didn't quite match the standards that lived inside my sensory mind.

Then, on a frigid February day, something happened that made it all right. Alex and I were down in the cellar topping barrels. The barrels lose wine by evaporation and sometimes by leaking, so every month they have to be refilled, or the wine will be exposed to air and breed vinegar bacteria. The wine that evaporates is known as "God's portion," and a good winemaker willingly gives God his due, as the evaporation concentrates the wine. Before topping each barrel, we would taste a sample to see how it was coming along. During a wine's first winter, it usually goes into a "dumb period," like adolescence, when the flavors become unpredictable and disappointing. Some of the wines we were topping were passing their second winter, however, and this is when they begin to show their maturity. This time, the wines that we were tasting were Merlots and Cabernets that had been languishing during the coldest months, and I didn't remember them as being anything special.

To remove each sample of wine from the bung hole on top of the barrel, Alex used a winemaker's tool called a *chantepleure*— a blown-glass pipette that draws out a glass's worth of wine at a time. *Chantepleure* means "sing or cry," which is what the winemaker will do, depending on how the wine tastes. Alex drew one sample after another and handed me the glass. I swirled each one, sniffed it, and filled my mouth before spitting it into the sink. One smelled a little like wet mittens. Another still had the aroma of *merde du chien* ("dog doo"), so aptly described by Peynaud as typical of young Merlot. There were hints of green pepper in some and tar in others. Then Alex took another sample, sniffed it, grinned, and handed me the glass. I put my nose over the wine and inhaled deeply. The tears started coursing down my face. I couldn't speak. This time, I was crying for joy. We had made the wine I had imagined us making. The wine of my dreams was in our barrel, in our winery. Whether it was the product of science or magic, no person could really be said to have made it. We created the conditions for it to happen, and then it all just came together on its own. Alex and I embraced each other with gratitude and relief.

12

Arrows in the Back

IN EVERY PIONEER STORY, THERE ARE CHALLENGES FROM nature. The snows, the gales, the parched earth, and the wild animals—it wouldn't be a pioneer story without them. In my case, even as a modern-day pioneer, these were the kinds of obstacles I expected. Obviously, my life in a vineyard in Cutchogue was not even close in terms of mortal danger to Laura Ingalls Wilder's life on the prairie in the 1870s. Nevertheless, the natural challenges were still there. When I pruned in the snow or picked in the rain, I was cold and uncomfortable, but I welcomed the feeling that I belonged out there, as much as the birds and the snakes. The same way that Laura Ingalls Wilder would tear off her bonnet so that she could feel the wind in her hair, I exposed my arms to the sun in a most unladylike way.

As we settled into life in the vineyard, our determination and enthusiasm carried us forward. With my husband, children, pets, grapevines, wine, employees, and visitors to attend to, I didn't have time to reflect on John Wickham's warning that pioneers get arrows in their backs. Where were the Indians? I didn't see them. But they were there.

It came as a shock to me to learn that Dr. Nelson Shaulis, the Cornell professor who had conducted an experiment in our vineyard from 1974 to 1979, had wanted our efforts to fail. In 1997 we got a letter from John Tomkins, the Cornell professor who had first told Alex about the North Fork as a place to grow vinifera grapes, that explained what had happened with Shaulis. Tomkins was the man who had brought vinifera table grapes to John Wickham's farm a few years before we saw them on our first trip to the North Fork in 1972. In his letter, he wrote, "I wish to comment about something which might affect the entire grape survival on Long Island."

Tomkins went on to describe how he had been Nelson Shaulis's first graduate student. Dr. Shaulis had gotten Tomkins a job doing grape research with him after World War II. "I thought Nelson was my friend," Tomkins wrote. He then described how Dr. Konstantin Frank, the vintner who had dared to plant vinifera vines in the Finger Lakes, had invited Cornell administrators to visit his vineyard. Dr. Shaulis had been telling everyone at Cornell that Dr. Frank's operation was "a disaster," but the dean of the School of Agriculture was impressed by it. Dr. Frank thought that some of his vinifera grapes would grow well on Long Island, and the dean agreed to fund an experiment there on Wickham's farm. Tomkins was the only Cornell pomologist who already traveled regularly to Long Island, so he was assigned to manage the study. "Trouble soon hit the fan," Tomkins wrote to us, "as Dr. Shaulis objected to my working with grapes there. He even threatened to resign. He applied for a job at Davis [California] but was turned down not because he was not qualified but because he had a speech impediment and might have problems speaking to growers."

Dr. Shaulis stayed at Cornell, and the Long Island experiment at Wickham's continued. Dr. Frank and Dr. Shaulis both went to visit it with Tomkins, but, as Tomkins put it, "never in the same time in the same car." At one point, Tomkins commented to Shaulis that he wished Dr. Frank had used more than one kind of rootstock on the vines he sent to Long Island. He was afraid that "if his vines died the dean might wonder what happened." In Tomkins's words, "Nelson replied—you are upset because you think the vines might die. He WENT ON TO SAY THAT HE WAS TERRIFIED THAT THE VINES MIGHT LIVE AND THRIVE THERE [Tomkins's emphasis]."

Because of this comment, Tomkins was now worried that Dr. Shaulis, whose job it was to advise New York grape growers of all viticultural problems, might not have sent (and in fact did not send) us his report warning of some serious rootstock problems in California. He wrote us, "I might sound rather bitter but did Nelson use his copy to alert the Vinifera growers in NY what was happening or did he use my report to him as toilet paper which he flushed down the drain?"

Tomkins tried to understand what would motivate his old friend Nelson Shaulis to want to subvert an industry. Was it jealousy, or Shaulis's own sense of impotence after a lifetime of hard work? Tomkins told us that Shaulis had worked on a book about wine grapes in New York for forty years. When he sent it to a publisher, it was returned to him, rejected. Tomkins speculated, "Perhaps the planting of Vinifera varieties was the death of his book. I strongly feel that he wanted a very cold winter to arrive and knock out the Vinifera grapes in the northeast."

Luckily for us, the cold winter that Nelson Shaulis wished for never came. We did have to deal with the legacy of his attitude, however, even after he retired. The results of the five-year Cornell study that was done in our vineyard were never published. Other Cornell agents who had worked with Shaulis whenever he came to Long Island continued to look for signs of problems in our vines—mites, funguses, crown gall, viruses, drought injury, and anything else. They would bring people by to observe whatever problems they found and alert the media about them, too. We had to deal with a steady stream of eager-beaver reporters who wanted to see the tomato galls that spotted a few leaves near the woods or Japanese beetles—none of them truly threatening—for themselves. When Dr. Shaulis's replacement upstate brought down a variety of experimental hybrids to be planted at the Cornell research farm on Long Island instead of vinifera—the only type of grape it made any sense to grow here—we got fed up and asked that they all stay out of our vineyard. We had had enough of their kind of help. (In recent years, however, Cornell Cooperative Extension has hired some very good extension agents who are trained in viticulture and are most helpful to Long Island's growers.)

The hybrid-vinifera grape wars continued without us. That masked bandit of the Finger Lakes, Walter Taylor, who had gotten his fans to publicly ink his name off hundreds of bottles of his wine, came down to Long Island to stir up more hoopla in his mask and cowboy hat. Back upstate, when he was cited for contempt of court for continuing to play games with the Taylor name, he obliged the court order to turn over all his offending literature to the Taylor Wine Company by bringing it in a manure spreader.

Meanwhile, the curmudgeonly Dr. Frank had persuaded Walter Taylor's winemaker, Hermann Wiemer, to plant some vinifera grapes. Walter didn't want vinifera planted in his own vineyard at Bully Hill; he was dedicated to hybrid grapes, and didn't like Dr. Frank any more than Dr. Shaulis did. Because of that, while he was still working for Walter, Hermann started his own vineyard on Seneca Lake, a warmer part of the Finger Lakes than the Bully Hill site. On Christmas Eve, while Hermann was visiting his family in Germany, he received a cable from Walter. Thinking it was a nice Christmas greeting, Hermann eagerly opened it only to learn that he had been summarily dismissed from Walter's employment.

We had bought our first vines from Hermann, so when we

got wind of this, we invited him down to Long Island to trade ideas about growing vinifera. To welcome Hermann, Alex made a big sign that he put on our tractor shed at the end of the drive-way, poking fun at Walter Taylor's motto, "Wine without water." Our sign read WINE WITHOUT WALTER.

While we weren't getting any help from Cornell, we certainly weren't getting any from the State Liquor Authority, either. The SLA maintained an elaborate system of keeping tabs on wine producers by requiring us to file twenty-five copies of a list of everything we had for sale every month, along with prices and quantities of wine per package. There were two versions of this, and they had to be printed on different-colored paper—yellow and salmon. Every month, Alex would religiously fill out the forms, take them to a printer to be copied on the right color paper, and send them off to the SLA. One day he received a notice in the mail that our winery license had been revoked for forty-five days. But what had we done wrong? He called the SLA and was transferred from one official to the next until finally he got someone who knew something. "You violated the filing regulations," the man told him. "How did I do that?" asked Alex. "I filled them out the same way I always do."

"You failed to specify the quantity of wine in each container," he was told.

It was true. Alex had put "12 bottles per case" at the top of the list of columns, and below that, he had used ditto marks because all of our wines were sold in twelve-bottle cases.

"Ditto marks are illegal," the official said. "That's why we are revoking your license until the next filing period."

In the end, Alex was able to negotiate a deal in which he wrote the SLA a mea culpa apologizing for the use of dittos. We got our license back and never used another ditto.

Our own intense enthusiasm served to shield us from the barbs of those who we could see were simply jealous of our success. From the very beginning, we took our dedication to making our farming venture work into the rest of the farming community. In the late seventies, after Alex stopped riding the tractor, I would often come in from the field to find him on the phone with members of the Farm Bureau who were trying to stop the construction of three nuclear power plants about six miles from our farm. The success of that effort led to his leadership in organizing an agricultural district so that farmers who committed themselves to a minimum of eight years of farming could get a tax abatement. He hit it off with Albin Pietrewicz (known as Al Patrick), a neighboring potato farmer with a round, red face and a quick laugh who teamed up with Alex. They paid calls on all the local farmers and persuaded many of them to preserve their farms.

All of Alex's organizing activities caught the eye of Governor Hugh Carey, who chose him to be an "honorable commissioner" on his Temporary Commission for Real Property Tax. To do this unpaid but distinguished job, Alex flew to Albany and other parts of the state for meetings and public hearings on new rules for real property tax. There was a move afoot to set a higher tax on farmers whose land was planted in more valuable crops. Again, we were pitted against the upstate grape industry, because our vinifera fruit was worth more per ton than upstate native or hybrid grapes. Alex was able to argue that there were too many factors involved besides price—yield, labor cost, land cost, higher risk—to make taxation based on crop valuation a valid concept. Again, his logic and rhetorical skills prevailed.

The commission took a great deal of Alex's time, but it did have two benefits. When Alex went to the State Liquor Authority office in the city to renew our winery license, he found himself seated at a desk across from a stone-faced official who pretended that he couldn't find our records. The man said, "Let me go see if I can find them elsewhere." He stood up, pulled out an empty desk drawer, and left the room. Alex realized that the empty drawer was a big hint for him to leave some kind of contribution for the official. Thinking fast, he put his Commissioner of Real Property Tax card in the drawer and closed it. When the man came back empty-handed and looked in the drawer, he immediately snapped to attention. "I think I can find your papers now, sir," he said, closing the drawer.

The second benefit of Alex's involvement in the commission came when his father, also named Alex Hargrave, an attorney who had been involved in politics upstate for years, called with a tale. "My secretary rang me to say that the governor was on the line for me," Alex's father said. "I picked up the phone and greeted the governor. There was a long pause, and then the governor said, 'Oops—I got the wrong Alex Hargrave!' "

During the eighties, Alex's role as the treasurer of the Hampton Day School embroiled us in more controversy. He and the rest of the board of directors agreed that problems with the school's current director were insoluble, and they acted to replace her. After half a year of intense meetings the search committee that Alex was part of hired a new director. This woman came with her own problems. By the time she was seen sunbathing in the nude in front of the school, Alex had removed himself from the board. It was too frustrating to work as hard and care as intensely as he did, and to see the effort go for naught. We didn't need to go looking for crises. Our lives in Cutchogue were about to be changed by something no pioneer on the prairie had to deal with—a modern regulatory agency.

I should have known that just when everything felt right, it was time for something to go wrong. On a chilly day in the late 1980s, an enforcer for the Department of Environmental Conservation (DEC) stood in our doorway in a leather jacket, clenching the muscles in his jaw. The expression on his face reminded me of a schoolmate of Zander's who liked to pull the legs off frogs just for fun.

The official was there to inspect our pesticide storage shed and check our pesticide application records. Alex held our pesticide applicator's license, so he showed him that the records were all in order, complete with labels and dates of application. The official took notes and went away.

A few days later Alex got a call from someone at Long Island's DEC office. He announced that we were in violation of pesticide-use regulations and that we owed a fine of six thousand dollars. If we signed a consent order, the fine would be reduced to three thousand dollars.

As with the purported violation of the law on ditto marks, we could not imagine what law we could possibly have violated. Alex knew the regulations and was scrupulously careful to follow them. It turned out that the DEC had traced some Ferbam, a common fungicide that we had bought, and found that its registration had lapsed at the time we bought it. The way we purchased pesticides was to call the salesman at the local farm supply company and ask for delivery of whatever we needed. The company brought it to us. Every pesticide was individually registered on a periodic basis by its manufacturer. In the case of this Ferbam, some employee of the manufacturer had let the registration lapse that period. It was just a matter of having neglected to pay a small fee to the DEC; the substance wasn't illegal. It was approved by the Environmental Protection Agency. Other brands of the same product that were identical had maintained their registration through this time. We never saw the package before it was delivered to us, and even if we had, there was no way of knowing whether its registration had lapsed or not.

When Alex went down to the DEC offices to speak with the

officials directly, they threatened to publicize our misuse of pesticides and ruin us. Incredulous, Alex turned on his heels and walked out, saying, "Let the party begin." How could he sign a consent order that was an admission of wrongdoing, when he hadn't done anything wrong that he could possibly have known about? Once a consent order has been signed, the next violation the DEC comes up with may be a criminal offense.

We learned very quickly that the DEC is funded by collecting fines from alleged violators. The jobs of the staff at the DEC depend on their pursuing violators—and who could object to that? It appeared though that there were some big violators who were politically untouchable, so that left the easy targets, like dumb farmers, for the DEC to go after.

The DEC enforcer had also checked the dates when we had sprayed the fungicides. Every year, we were issued recommendations for pesticide use that Cornell University had been charged by government agencies to write. All the farmers used these guidelines for spraying, and our spray dates coincided with these recommendations. The enforcer was a real sleuth, however; he figured out that the Cornell recommendations did not coincide with those on the pesticide label. So there was another claimed violation. Every day we failed to sign the consent order, the ticker ran until the DEC asserted that we owed in excess of ninety thousand dollars.

Because we were not at fault, we decided not to sign the consent order under any circumstances. Alex called Cornell University to get some support from them, since it was the university's recommendations that were getting us into trouble, but they would not take Alex's call.

The next day, Alex and I were working in the wine cellar when we got a message from one of our employees who was at our retail outlet on the highway. "There was a man here from the DEC," she said. "He wants to find you, but I said you were out.

He wanted me to give you some papers, but I said I didn't know when I'd see you."

It was sheer luck that one of us hadn't been there when the DEC official came. He was trying to serve us a summons. The DEC has its own regulatory quasi court, and we weren't optimistic about the outcome if our case was heard there. After a powwow with our lawyer that went into the night, we decided that we could not allow the DEC to serve us that summons. The agency had thirty days to serve us before we could take the case to the regular courts and get relief, and two-thirds of that time had passed before we decided to hide out.

Now I knew what it felt like to be a fugitive. Billy the Kid didn't have anything on us. The DEC people couldn't trespass onto our farm, but they could sit at the end of our road and wait for us to come out. If the DEC made good on its promise to tell the press that we were outlaws, it would ruin us. No one would care about the facts; they would just believe that our wines were tainted with illegal pesticides. I imagined a reporter calling and asking, "Tell me, exactly how did you put poison in your wines?"

Alex and I decided that we had to get our children out of there. We arranged for them to stay with their teacher in Montauk until the end of the service period. We told them not to call us, fearing that the call would be traced; in fact, we took no calls at all during that time.

We were able to keep working in the winery, which was directly adjacent to our house, but we had to rely on one of our employees to bring us messages and supplies. Repeatedly, the DEC tried to serve the summons. At night we kept the lights out. If we passed a window, we crept under it so as not to be seen. It was hard to eat, hard to sleep, harder still to believe that we were hiding from the law because some secretary forgot to mail fifty dollars to the DEC.

When the time for service elapsed, we went to the state

supreme court to get an order to stop the DEC from proceeding. There, the judge looked at the DEC's procedural irregularities as well as the substance of the claims against us. Finding both irregularities and lack of substance, he appeared to be incredulous that the case had gone so far as to land before him, and he ruled in our favor. The DEC did not pursue the matter against us.

Just because we won the shootout didn't mean we felt victorious. What was the meaning of all our labors if a bunch of petty bureaucrats could threaten to ruin us? Was it for this that we had risked everything and put our hearts on the line, hoping to build a life for our children and ourselves? Where did our love of nature and our desire for an honest day's work fit into this equation? We decided to quit.

We had a meeting planned with our partner, Bill Chapin, who still owned almost half of our corporation. Bill had been an ideal partner—interested but undemanding. He had called the meeting to discuss another investment he wanted to make in Florida with Jack Gross, a man who was also his attorney. The meeting was in Rochester, Alex and Bill's home town. On the long drive up there Alex and I talked about how we would present the idea of quitting to Bill. It was a pretty straightforward matter of putting the vineyard on the market, but his aunt still owned one of our buildings, and there were details that would have to be worked out concerning equity because Alex and I had not taken regular salaries since the vineyard started.

As I recall, we met in a private room at a country club. Alex and I were surprised that Mr. Gross (whom we had not previously met) was there with Bill, but the meeting began well enough. We all shook hands, settled into chintz-covered chairs, and said friendly things to one another. Having lulled us into submission, Mr. Gross proceeded to unroll plans for an elaborate

Florida condo that Bill had designed. We politely admired the architectural details and ambitious expanse of the development, until Mr. Gross abruptly changed the topic of conversation. He glowered at us and said, "You know, your vineyard has been a terrible investment for Bill. Nothing in your business plan has come true. The million dollars that Bill inherited is all gone, and it's your fault."

Mr. Gross might as well have punched us in the face. We knew that Bill had lost money, but it was because he had pledged his stocks to the bank and gotten stung when his stocks lost value and the bank called the loan. It was true that we hadn't been able to pay him—or ourselves—what we had projected back in 1973, but we had built a valuable asset with a strong reputation, which protected his investment more than the stock market had.

Bill sat there while Mr. Gross rubbed it in some more. Mr. Gross didn't even know us, and here he was assassinating our character, telling us that we had to buy out our partner so that he could build a bunch of condos! I felt like a kid whose best friend had just sold out to the school bully. Both Alex and I were speechless. We were so close to tears that we just said, "We'll get back to you," and left the room.

We couldn't sell the vineyard. Not now. Didn't all our awards and accolades count for something? What about the life we had made for ourselves with our children. No! We would not be insulted! If Bill had come to us alone and said, "Gee, guys, this vineyard thing just isn't working; can we figure out a way to exit gracefully?" we would have agreed with him. But Mr. Gross had no right to insult us the way he had.

We went back to the bank, refinanced everything we had, and worked out a deal with Bill that bought out his interest in the vineyard. After the real estate closing, instead of going home, Alex and I drove to the sound and sat together on the rocky shore. We watched the gulls bickering over a piece of dead fish. The tide was coming in, and the small waves made a soft gushing sound as they pushed against the shore. Holding hands, we gulped in the fresh sea air. I thought of sampling wine with the *chantepleure*. Should we sing or should we cry? I wondered. The vineyard was now all ours. And so, we sang.

13

Just Plain Folks

IT WAS SPRINGTIME AGAIN ON THE NORTH FORK. AS I WORKED in our Chardonnay vines along the highway, the sounds that I heard told me that the frontier was settled. My crew and I could no longer hear birds singing as they built their nests. We could no longer hear each other speaking while we worked those rows; there were too many tractor-trailers on the highway, hauling building equipment to the farm fields where second homes were replacing potatoes as a cash crop. If it hadn't been exactly like the Wild West before, now it was very much like Buffalo Bill's Wild West Show.

By 1988 there were thirty vineyards whose thirteen hundred acres of vines produced grapes—all vinifera—for twelve wineries. The vines that we had planted in 1973 and 1974 were not venerable old clones; they were suffering from the excess vigor that had resulted from their being grafted onto a hybrid rootstock, and many had succumbed to crown gall or dead arm disease. We had planted other vines on stock from California, but they had all arrived with problems, and none of them had lasted even as well as the first vines we planted. We were down to about twenty-five acres of vines on our eighty-four acres of land, and we didn't have the ten thousand dollars an acre we would need to replant the rest. Besides, we heard rumors that the nurseries in California were having problems with a mysterious disease known as "black goo," and we didn't want to get stuck with that. Meanwhile, our wines were doing well, but we needed more fruit to meet the demand for them.

One of the songs my mind kept playing was the Rolling Stones anthem "You can't always get what you want, but you get what you need." The Stones were right. In the early summer of 1988 Alex received a call from an agent for a New York developer who had formed a partnership with a local speculator to start a vineyard that was about a third of a mile from our farm. The partner had gotten into trouble, and the developer was unhappy with how much he had been paying a manager to tend his vineyard. There were fifty acres of Chardonnay, Pinot Noir, Cabernet Sauvignon, Cabernet Franc, and Merlot, all five years old and bearing the beginnings of a crop. Alex negotiated what looked like a very good deal—basically, an exchange of our work for their crop. We signed a lease and immediately went into the fields that we called "King Tut" (after the Tuthill family, which had owned them for almost three centuries) to catch up on the work that had been neglected there.

This was a farm that I had admired for years. When we first came to town I used to buy eggs from the aging Tuthill brothers, Stanley and Leslie. Stanley liked my son, Zander, and gave him several Indian arrowheads that he had found on his land. He told me that before they used tractors and disks to cultivate, he would find Indian artifacts in the field almost every time he worked there. Now, he imagined, they were all chopped up by heavy farm equipment.

The two brothers had lived alone on the farm for many years, after their parents died and their sisters got married and moved away. Stanley (whose nickname was "Sparky") was sharp and outgoing. He was clearly in charge of the farm. Leslie was scrawny and very, very shy. He was in charge of the chicken business, and after I had been buying eggs from him for a couple of years, he also let me buy cream from the cow he kept tethered in the back yard. This cream was so thick that if I turned the quart jar of it upside down, it stayed in place.

Every day at nine A.M., John Wickham would come for coffee with Stanley, Leslie, and one of their cousins, John Tuthill. Sometimes they were joined by Ed Gilles, a commercial pilot who was the only one of the group to travel off the North Fork on a regular basis. Stanley had seldom left the North Fork, and I'm not sure if Leslie had at all. I liked to join them to learn what was going on in town, and to hear about the old days on the North Fork.

"When I was a boy," Stanley told me, "we never ate anything that wasn't blemished or partly rotten. We grew all our own food, and if an apple or a squash or a potato was perfect, it would be sent to market. I wanted to eat store-bought bread more than anything, but we couldn't afford such a luxury. I had to eat the bread that my mother made."

Stanley's father assigned him and Leslie each a row of potatoes that was their responsibility, and any spending money they had would have to come from that row. One year, Stanley said, his crop failed. He had been counting on the money from it to buy a winter coat because he had outgrown his old one. That winter, he went around with his wrists sticking out of his tattered old jacket.

One day when I stopped in, I saw him making sixty sandwiches—baloney on Wonder Bread. He said that he was going to freeze them so that he and Leslie could have a month's worth of lunches ready when they needed them. He got his wish for store-bought bread, I thought. He told me that the one thing he wanted in life was a Rolex watch, and he got that, too. In the mid-eighties, he sold his farm to the developer, with a lifetime tenancy for himself and Leslie. He bought himself a classic Rolex with part of the money.

Shortly after that, when I went to buy eggs there, I learned that Stanley had almost lost his watch. He had gone out to work in his vegetable patch, and to protect his watch from getting dirty, he had left it on the back porch next to the strongbox that held the egg money. A priest and two nuns came to the back door of the Tuthills' farmhouse, where customers usually pulled in to buy eggs. They weren't really God's anointed; they were robbers in disguise. While Leslie spoke with the priest, the nuns went around him into the house and took the money out of the strongbox. Leslie wasn't aware of what had happened. When Stanley came inside and discovered that the money was missing, he couldn't believe that his Rolex was still sitting there on the shelf.

While I loved the Tuthill brothers and I loved their farm, I wasn't sure that our crew could get the work done on an additional fifty acres of vines, when we could barely manage our own acres. We had only two full-time helpers in the field besides myself, and we were the same crew that also worked in the winery with Alex. We added part-timers for tying vines, weeding, and harvesting, but those kinds of workers were unreliable. Furthermore, after the charade with the DEC, we decided not to use any "restricted" pesticides. That ruled out using chemical weed killers, which meant that our mechanical cultivation would have to be even more timely. Notwithstanding my reservations, the opportunity to get more grapes was too promising to turn down.

One thing that made it a little easier for me to commit myself to farming the additional fifty acres was that, except for vacations, both my children were in boarding school after 1987. Alex and I had both been to boarding schools and had loved the experience. We agreed that being teenagers away from home had made it easier to become independent without creating friction with our parents, and we wanted the same independence for our children. Furthermore, we were frustrated by the limitations of our local schools. We had moved our kids into different schools, public and private, through their elementary years.

Although our original plan for the vineyard was centered very much on having our children with us, we found that it was unrealistic to think that they could participate much in vineyard work while getting their schoolwork done as well. The vineyard was so far away from their friends and sports activities that we had to spend large chunks of time transporting them when we needed to be working ourselves. When Anne was thirteen we enrolled her in a school in Virginia. As for Zander, we sent him away the same year Anne went away, when he was a ten-year-old sixth-grader, because of something that happened to him while visiting a classmate.

Zander's grade school divided students into sections that encompassed three years. The idea was that the younger students would learn from the older ones, and besides, the enrollment in his school was too small to do it any other way. The result of this was that nine-year-old Zander was in the "upper school" with his sister and eight other students, aged ten to fourteen. Zander had skipped second grade, and to fit into this older group, he insisted on listening to heavy metal music and wearing T-shirts with heavy metal band logos that were popular among the boys.

Clyde, the oldest in the class, was supposed to be taking lithium for his emotional problems. No one knew that he had stopped taking it because of the side effects.

One November weekend, Clyde invited the boys in the class to his birthday party. Because Alex and I had plans to be in New York City that weekend, Zander was invited to sleep over at Clyde's. On Sunday, when Alex went to pick Zander up, he noticed that Zander was not his boisterous self. "Did you have a good time?" he asked him. "Sure, Dad," Zander replied. "It was fun." They drove on, but after several miles had passed, Zander said, "Dad, it wasn't fun. Clyde tried to kill me."

Alex stopped the car and listened to Zander's story. "Clyde got a bowie knife from his mother for his birthday. Before the other boys came Clyde told me to come play outside with him. He made me stand against the wall while he threw the knife around me. Then he got his BB gun and fired BBs at me. After that he got his brother's pistol. He said it was loaded, and he put it to my head and pulled the trigger. Nothing happened. He pulled the trigger again. I tried to run away, but when I got to the end of the driveway, I realized I didn't know where to go. The other boys came for the party, and it was okay after that. I don't know where Clyde's mother was."

Alex spun the car around and drove back to the police department. He told them Zander's story and drove Zander home. We later learned that the police confiscated the knife, the pistol, and the BB gun.

At school on Monday morning we told Zander's teacher what had happened and asked that she keep an eye on Clyde. She said that he was a boy who had had a hard life, and he needed another chance. Zander came home from school and said that while he was waiting for his ride home Clyde had put his hand, formed like a gun, up to Zander's head and said, "Click. Click."

We pulled Zander out of the school, and Anne, too. She finished the term in public school, and Zander went to a parochial school. At the end of the first week in his new school, he said, "Phew. I'm glad I don't have to like heavy metal music anymore."

At the end of that year, Zander graduated from the parochial school, which ran only through fifth grade. He had already tried public school and been bored there, so we decided to send him to boarding school like his sister. He called us the first night of school. "How's it going, Z?" I asked. "Good," he said. "There's only one thing. Guess who's here?"

Clyde.

Whether or not we made the right choice in leaving Zander there after we called his new school and informed the staff about Clyde could be endlessly debated. Zander did well, and Clyde eventually left.

The upshot of having the kids away during the school year was that I had more time to work. And now, with King Tut, I had more work to do.

Back in the days when Alex was studying Chinese, I had memorized one of Chairman Mao's sayings: "If you don't exercise, you won't build up strength." I bought a bright blue running outfit, and started running up and down our farm road every morning. Zeus ran with me. He would go tearing around the borders of the vineyard, his tail high and his ears flapping. Once he had made the rounds of the boundaries, he would check out the snake and gopher holes and other things of interest to a dog. In the time it took me to run from the house to the tracks, he would have made a complete survey of everything. I was a terrible runner—my knees were bad and I stumbled breathlessly over the potholes in the dirt road. Even so, I loved being outdoors on the farm as the sun came up. I would see hawks soaring above the fields, and deer venturing out of the woods. This was something new for me—time alone. There wasn't much of it, but it was exhilarating.

Another sort of exercise—putting each wine together before bottling it—was an exercise of the senses that I also enjoyed. Once we had a little experience doing it, it became fun and very stimulating. Alex and I had learned everything we knew about wine together, so we shared a wine-tasting vocabulary, and we usually liked the same style of wine. The sensation of smelling wine goes to a part of the brain that is nonverbal, which is why it is so difficult to put the wine-tasting experience into words. We made an effort to do it anyway because we needed to communicate with each other about the minute nuances of each wine. If a three-thousand-gallon tank of Merlot is put into sixty-gallon barrels, after a remarkably short time in oak there will be fifty batches that are each now subtly unique. We had to be able to notice and describe those differences in order to make decisions about which barrels should go together.

Besides putting together different batches of the same kind of wine, we would sometimes blend wines made from different grapes. When the name of a grape (like Chardonnay) is on the label, the wine becomes a "varietal" wine, and by federal law it must have at least seventy-five percent of the named grape variety in the bottle. That leaves twenty-five percent leeway for the winemaker to blend in another kind of wine. Say a Sauvignon Blanc is too hard and grassy. Put in just five percent of Riesling, and suddenly the wine has delicate grace notes. Maybe a Merlot is too soft. A judicious jot of Cabernet Sauvignon will give it structure.

The way Alex and I worked on a blend was that he would assemble several different options, using a graduated cylinder to measure various proportions of the wines being used. Once he had them labeled A, B, C, or whatever, he would call me into the lab to taste them. I would never know exactly what he had put together, so my tasting was purely for quality. I would take each sample in its glass, swirl it to release its aroma, and make my most important judgment in the first split second of sniffing. If I had to keep going back to the glass to smell it again, my nose would get acclimated to it, and all the other aromas in the room would start to throw me off.

Alex and I would hover over the lab bench, resting on high stools while we sniffed. "B smells like peanuts," I might say, or, "It's too empty." Alex didn't want to hear about what I didn't like; he wanted to hear what I liked, but I usually got to that next. "Ooh, I just got some black currants," I might say, and then he would take the glass, sniff deeply, and come back with "Black currants with a bit of fig."

After we both tried to describe what we smelled, we would put each wine into our mouths and let it roll around our tongues and palates, sucking air over it at the same time. At the end of each slurp, I would slosh it around my mouth like mouthwash and then spit it into the sink. We swallowed a bit of each sample to test the aftertaste—probably the most important aspect of a wine in the long run.

After we sniffed and tasted all the batches, the real work would begin. "I like the nose in A, but not the finish," I might say. If B had a better finish, maybe we could try something in between the two. A single percent of some wines could noticeably alter a blend.

Often, the best wine was also the most costly to make. If one barrel out of twenty was exceptional, would we bottle it as a reserve and have only twenty cases, or would we use it to improve the rest of the wine, making something good but not as good as the best? Since Alex was the one who juggled the wine from tank to tank or barrel, he was aware of which of his blends would make the largest bottling and potentially earn us the most money. It was a business after all. Sometimes he would lobby for the money-making blend, but most often he agreed with me that our reputation rested on making the best wine we could.

Blending wine wasn't something we did every day. I spent an increasing amount of time giving tours and selling wine to visitors. There was no such thing as a weekend off for me; Saturday and Sunday were the busiest days at the New Barn. Those were the days that would make or break us. Most of the people who visited the vineyard were nice. Often, visitors would be lively and interesting. I actually loved telling them stories about winemaking and showing them how to taste wine. But the hospitality of wine tasting was a far cry from being outside in the vineyard. I guess the pioneers on the old frontier had to hawk their produce, too. It was part of the business that I had not thought much about before we started. The trouble with it was that after a day of trying to give our customers a good experience, I had no more left to give my family. The only way I could restore my equilibrium was to cook. The physical act of chopping or stirring would set me straight again.

One day, after I had spent eight hours trying to overcome the skepticism or ennui of flocks of wine tasters, I walked into the house ready to start making dinner and smelled wet paint. What was Alex up to? I wondered. Before I went to investigate, I washed up and put a hot cloth on my face. My jaw ached as much as it had before we started the vineyard. This time it ached from smiling all day at the visitors who came to taste our wine. "Welcome to Hargrave Vineyard," I'd say, just to make sure they weren't lost. To make them feel comfortable I'd ask, "Have you been here before? What kind of wines do you like?" The answer was often obvious as they entered the building clutching a beer can. One fashionably coiffed woman removed the chewing gum from her mouth with great care and balanced it between the ring and middle fingers of her right hand as she reached for the glass I offered her. Families had swarmed in the door, and while the children swooped around the tiled vestibule, shrieking, their parents pretended that the kids belonged with someone else. A newlywed couple watching this had their first conversation about children and discovered, as they waited at the register to buy wine, that he wanted five children and she wanted none.

When I got to Alex's study I found that he had painted the walls a glowing saffron yellow, the same color as a Buddhist monk's robes. "Nice color," I said to him. He nodded. He was absorbed in his reading. Looking up at me, he told me that he had been to Southampton that morning, making a delivery of wine that we had donated to a local charity. The charity had called the night before as if it were a dire emergency that none of the staff had managed to come up with free wine for this five-hundred-dollar-a-plate dinner. Could we please bring five cases, and could we deliver them by ten A.M. Saturday? Even though we had just finished our Friday deliveries to the Hamptons, Alex had agreed to bring the wine himself. He related to me how he had arrived at the seaside mansion owned by an anchorman of one of the New York network newscasts. Caterers were bustling around a tent, and the newscaster himself was giving directions. This man looked at the very substantial, six-foot-six-inch Honorable Alex Hargrave, who was wheeling the wine down the driveway. The newscaster pointed to a pile of party supplies and said, "Put it there, boy."

Alex looked the man in the eye, turned, and left the cases on the designated pile. He bought the paint on the way home. It was the beginning of a long meditation.

Soon after this, we finished work on part of the addition to our house that had been abandoned after Hurricane Gloria destroyed our cash flow. The upstairs room that was meant to be a master bedroom had a view of the whole vineyard, and we agreed that it would be wasted as a place for sleeping. Alex furnished it as a new study for himself, and it became his private retreat. The mayor of Rochester's desk couldn't accommodate a computer, so he used his grandmother's dining room table—the same one where, in 1972, we had eaten the crème brûlée made by Rose, the German cook, who had wept when she heard our plans to start a vineyard. Under the picture window, where he liked to watch the blackbirds chase the hawks through his binoculars, he placed an antique American Empire settee. It had hand-carved ball-and-claw feet that made it look like a subdued animal, resting until it had a chance to grab its prey. Alex covered the walls with shelves for books and CDs, but he left space to hang a map charting the constellations of the stars, and an etching of a nude woman with wild hair, whose arms and splayed legs reached toward the viewer in an inescapable embrace.

Alex continued to manage the farm. He still put together the wines, and during harvest it was he who stayed up all night to mind the press. But after the DEC prosecuted us, his focus turned back to his own private intellectual pursuits. Besides the poetry that he loved, he studied the ancients again. Virgil and Confucius were his old friends. Moving into the eighteenth century, Alex read Carl Linnaeus's taxonomy of plants, the *Systema Naturae,* and Erasmus Darwin's *Zoonomia.* Visiting him in his study, I picked up Linnaeus, who believed that the study of nature would reveal the divine order of God's creation, and read a description that he wrote in 1729:

The flowers' leaves . . . serve as bridal beds which the Creator has so gloriously arranged, adorned with such noble bed curtains, and perfumed with so many soft scents that the bridegroom with his bride might there celebrate their nuptials with so much the greater solemnity.

Alex followed the naturalists into the nineteenth century, inevitably delving into the works of Erasmus Darwin's grandson, Charles Darwin, who wrote so famously about the survival of the fittest. Alex also investigated the language of the flowers and returned to the theories of cataclysmic evolution that had interested him when he was stuck in a body cast so many years before. The librarians at the Suffolk County Library system told him that only one person borrowed more books than he did—a woman who took out trashy novels. Alex made requests for things like the journals of wilderness priests, Tasmanian dictionaries, and popular fiction from the 1830s.

I couldn't reliably say that Alex's deeply consuming studies were a substitute for some fundamental need to be in touch with nature. I myself felt that need, but it was satisfied by my jogging and by my working outside in every season. With the kids away I had some time to read at night, too, but my choices were less far-flung. I liked to read biographies, especially ones centered on the American Civil War era, such as *The Personal Memoirs of Julia Dent Grant,* General Grant's wife, who was clairvoyant, and the *Diary of a Slave Girl,* who wasn't. Sometimes, I'd read something that was a complete waste of time but immensely enjoyable, like a Dick Francis mystery. My reading was for relaxation.

By the time we had been on the North Fork for fifteen years, the cycle of the vine was a matter of routine, but even so, it was difficult for us to keep up with the work. Farming the vineyard at King Tut was always a source of trouble. Before we took over, the vines that had been trellised according to Cornell recommendations at a height of six feet had died back to where vinifera vines want to grow. This was at a height of about four feet, which was where our own vines had been trained all along. My crew and I had broken at least a pair of loppers a week cutting out the iron-hard deadwood. Renewing the vines doubled the work load there. We had two year-round employees, Betty Cibulski and Mark Terry, who started working for us around 1985 and who helped us both in the field and in the cellar. Betty also helped me in the store and, occasionally, in the house. I have never met anyone more considerate, energetic, or reliable than Betty. If I needed someone to come early or stay late, she would be there, cheerfully indefatigable. Because of the size of King

Tut, we often worked late. In the spring, when we needed to tie all the vines to the trellis before they budded out too far, we had to hire extra people. I would put an ad in the local paper and cross my fingers that whoever showed up to work would be better than the group from a halfway house I once unwittingly hired, who had threatened to beat one another with chains. Most of the help we found was okay, but there was one helper we hired whose situation had a blood-curdling effect on all of us.

I was in Alex's office when the phone rang. A man was calling in response to my ad. "I am looking for work for my mother, Rani," he said in a voice with an Indian accent. "She just arrived from Bangladesh. She lives in a trailer near you. Rani does not speak English, but she has papers and she can work very hard."

"Have her here at eight o'clock tomorrow," I told him. "We'll give her a try. Make sure she brings something to eat and drink with her."

The next day, Rani was dropped off at the vineyard on time. She was a weary-looking woman of about thirty, wearing traditional pants and a beat-up sweatshirt. Her solemn expression and attentive gaze indicated that she wanted very much to learn how to tie the vines properly. Her fingers were nimble, and it didn't take her long to understand how to shape the plants on the trellis. When it came time for lunch, when the other tiers went to their cars, Rani squatted near them, watching them eat. She had not brought any food after all. Several minutes passed and then she got up and, with her hands in a begging posture, indicated that she was hungry. There were four or five other women there, and they all found some food to share with her. She devoured it ravenously.

The next day the same thing happened, and the next. We all started bringing extra food for her. The second week of tying came and Rani still was begging for her lunch and ravenous when she ate it. I had thought that once she got paid, she would bring food. It wasn't right for my employees to feel obliged to feed her, so I took her to a small local grocery store. Looking completely overwhelmed, she picked out a bag of rice, eight lemons, a head of broccoli, and a box of doughnuts. When I drove her back to where she was staying I discovered that she was camping in a tiny old trailer that was hidden behind some old barns on the Kaloski property across the street. Why would Mike rent her the trailer, I wondered, when she wasn't working for him? It didn't seem like Mike. Oh, well, I thought, maybe he wanted a little cash. I left Rani standing outside the trailer with her groceries, and wondered how she would cook the rice in that run-down space.

On Thursday, the day after payday, Rani came to work covered in bruises and barely able to walk. She tried to hide her condition, but finally made gestures to tell me what had happened. It wasn't hard to understand what she meant when she made the shape of a man with her hands, and then beat herself with closed fists. Putting her hands in her pockets, she pretended to take out money and acted as if she were giving it to someone. Clearly, her "son" had come to take her paycheck. When he discovered that it was short the amount we had deducted for groceries, he had beaten her. Little by little her story emerged. Pointing to the red dot on her forehead, she indicated that she had been married. Her arms crossed over her chest, eyes closed, told me that her husband was dead. She cradled her arms and held up four fingers to indicate that she had four children. I was able to figure out the rest. Her husband had died recently, leaving her with four children and no way to support them. Someone who knew that she was destitute in Bangladesh had offered to bring her to America and find work for her. He had promised to take care of her and to send for her children in a few months. In fact, she became the victim of a slave organization, working for this man who had pretended to be her son. He had somehow discovered

Mike's trailer and put her there with no electricity or water. Mike knew nothing about it. Rani was alone and well hidden when the slaver came to beat her.

How could we get her out of this situation? What could we do? I wanted to make sure I understood her story correctly, so I put in a call to the Bangladesh consulate in Manhattan, thinking that at least the officials there would speak her language. When I explained Rani's situation, they said there was nothing they could do, but they would agree to speak with her. When I put her on the phone, she spoke briefly and hung up. I called them back and learned that she was too ashamed and frightened to tell them the truth.

Alex then called a county agency to try to get some help for Rani, but before anyone arrived, she disappeared. Rani's situation showed us the entrails of labor in our free world.

We had another worker, Jim, who left me shivering just to think of what his sad mind drove him to do one day on his lunch break. I had gone out on deliveries that day and was returning in our van when I was stopped in traffic just two miles from home, in Mattituck. There was a score of county police cars, lights flashing, clustered around an office building on the main road. Whatever was happening had to be serious, because our town had its own police force and seldom called on the county's. I saw ambulances, too, but no wrecked cars.

I idly listened to the radio as I sat there, and was daydreaming when the news came on. "A deranged young man has taken his psychologist hostage in Mattituck," I heard the reporter say. "He is armed with a rifle, and the hostage squad is negotiating with him."

So that's who those cops are, I thought. The hostage squad. Pretty unusual for my town. I wonder who the guy is?

When I got home, I found out that Jim had gone out at lunchtime and had never come back. He was the villain. Nice, quiet Jim. Mark and Betty, who had been working with him all morning, hadn't noticed anything strange about him. What we didn't know was that Jim had been in counseling since his parents had died when he was in high school and he had tried to commit suicide. We learned from the papers that he had become attached to his counselor, and was distraught when she told him that she was planning to go to South America to adopt a baby. He said he never meant to hurt her, but he had gone out that day and bought a gun, just to scare her so that she would agree not to leave him. When the hostage squad came he was frightened and threatened them, too, compounding the charges against him.

Jim went to jail for a year, and then, after he was released, he violated the counselor's order of protection by calling her on the phone. He was incarcerated again, and we lost track of him. I never believed that Jim would hurt a flea. The hurt was all inside of him. Still, when Betty and Mark pointed out how Jim used to work in subfreezing weather wearing a T-shirt, old jeans, and sneakers, I had to admit that maybe ice ran in his veins. It left a dead spot inside of me, to think that a person who functioned as well as Jim was at the brink all the while.

We had other disturbing experiences with employees—one who went on a three-day binge and returned with a broken needle stuck in his vein, another who was seen soliciting tricks behind the local Kmart and was later found dead behind a trailer in Florida. We hired a young man who failed to tell us that he was deaf, so when Alex explained to him how to shift the tractor gears he pretended that he understood, but he stripped them. Another promising fellow, a doctoral candidate in biology who wanted time out after spending two years working on the wrong genetic material—his professor's fault—could not bring himself to make pruning cuts. Still another had dropped out of MIT because it was just too savagely difficult. He was a good worker but developed an allergy to something in the wine cellar. Maybe to wine.

One employee whom I especially liked—a woman who was smart and sassy—was always so hung over that she came to work in her pajamas and began to store her clothing in the labeling room; she also took naps on the cases stacked in the back room. When we were trying to figure out what to do about her, my daughter said, "Ma, you hire the handicapped."

Because it was so hard to find good help, it was especially devastating when Betty, who had become like a sister to me, was injured while she was labeling wine. It was Christmastime, and she and I had started work at seven A.M. so that we could get a large special order ready. She worked the labeler while I assembled gift boxes. Another helper, Liz, organized the order. Our labeling machine, made by a company called Labelette, had been a source of frustration since we had bought it in the late seventies. To work it, the operator had to lay a bottle horizontally on rollers while pulling each label down from a vertical stack, using a lever with the other hand. A separate roller passed through a tub of glue that was heated to 250 degrees and then transferred the glue to each label. The machine required constant adjustment, and with its moving parts only Alex, Betty, one other part-time employee, and I knew how to operate it.

I was tucking the flaps of a box together when I heard a crash. There was another, grinding sort of noise and Betty cried out. She had dropped a bottle, and, in trying to catch it, had caught her fingers in the rollers so that they were immersed in the hot glue. Liz called 911 while I unplugged the labeler and tried to find a wrench that would let me move the rollers. Failing that, I went next door to get some ice for Betty's fingers while we waited for the rescue squad. All the while, two fingers on Betty's left hand were cooking in the hot glue.

A policeman reached us first. He used a crowbar to pry the

rollers apart and extract Betty's fingers. A minute later the Cut-chogue Volunteer Fire Department arrived with about thirty men, an ambulance, and a few fire trucks for good measure. They strapped Betty onto a stretcher and took her off to the hospital.

Betty had two fingertips amputated. She was out for several months for healing and therapy. Immediately after the accident I had to get back to work on the labeler. The orders still had to go out. For the next several months, since we didn't want our other employees using the Labelette, I did all the labeling, working extra hours in the morning and night as my mind re-played the accident. I dreamed of being immersed in huge vats of hot glue, in scenes like the hell of Hieronymus Bosch. Some-times, I cried for Betty's pain as I worked the machine alone. I could hear my tears spitting as they fell into the hot glue.

A short while after she returned to work, with fingers still hypersensitive to touch and to cold, Betty sued the Labelette company for damages. Workman's compensation laws prevented her from suing us, but Labelette sued us itself, claiming that it was our fault that the machine had no safety guards. Inevitably, the two suits were joined together.

It took several years for the suit to come to trial. In the meantime, Betty still worked as well for us as ever, but she un-derstandably curtailed her hours and activities. I missed the ca-maraderie that we had always had—we were still friendly, but our relationship changed. I dreaded being in court and having to defend myself against the accusation of having contributed to Betty's injury. I hated that machine. For years we had hurled in-sults at it, and now it had had its revenge. Changing to a safer machine—one that used pressure-sensitive labels—meant dis-carding thousands of dollars' worth of labels, and changing the label design as well. It took us several months after the accident to make the change to a new machine.

It was a warm day five years after the accident when I found

myself sitting on the witness stand in Suffolk County Supreme
Court. It was humiliating. The lawyer for Labelette made me
feel as if I were some kind of criminal, insisting that I answer his
accusatory questions with only yes or no. I thought his questions
required more explanation, so I disregarded his instructions. He
kept making objections, like Perry Mason, but the judge finally
let me speak.

Labelette had hired engineers to make drawings of the labeler. They showed the jury detailed drawings of the rollers, the
glue pot, the levers, and the lack of safety guards. Then, their attorney placed in evidence a manual for the machine that included a diagram of the rollers with safety guards. That put the
liability squarely on us. What that lawyer hadn't reckoned with
was that Alex had gone into our storage space in the barn above
the winery and found the original manual that had come with
the machine. It showed no guards at all. He had also found the
receipt for maintenance done on the machine by Labelette me-
chanics, several years after we bought it. The receipt included
notations about warning labels that had been affixed to the machine, but made no mention of the addition of guards because
there were none. When the attorney for Labelette saw our au-
thentic manual, he immediately stopped the trial and offered
Betty a substantial settlement.

Safety on a farm is an issue. We knew it before Betty's accident, and we were all the more aware of it after her accident.
Nevertheless, when an inspector from the Occupational Safety
and Health Administration (OSHA) arrived to inspect our
premises, we were tired of the whole subject. Besides, we had
too few employees to be under OSHA's purview. Still, knowing
that he had no legal obligation to do so, Alex decided that it
would be easier to comply with the agency than to argue, so he
led the inspector into our winery building. Being a fanatic about
cleanliness down there, he wasn't worried about what she would

find. But he had forgotten that at the foot of the stairs leading into the cellar there was a large black rubber mat that was losing its little rubber prongs. A few of them were scattered on the floor near the mat and, wouldn't you know it, the inspector noticed them. Looking aghast and pointing at them accusingly, she exclaimed, "Rat feces!" Alex was silent for a moment. Then, bending his large frame over, he picked up a prong and put it in his mouth.

14

Scalped

FOR ALL THE DISASTROUS THINGS THAT WENT ON AROUND US in the second decade of the vineyard, there were many good things that happened, too. The good things centered on the wine, which was a source of great satisfaction. We organized symposiums and wine dinners with other Long Island vintners; we traveled to California, France, and Italy, always enjoying wonderful food, fabulous wines, and the company of generous winemakers. I could never forget the sparkle of a Burgundian *crémant* served to us in Mâcon by the Vincent family of Château Fuissé as we sat at their picnic table in the sunlight of an early spring afternoon. What divine chef invented the *poire Bacchus,* a dessert of wine-poached pears trapped under a cage made of spun sugar, that we ate across the street from the train station in Bordeaux? What could replace standing by the shore of the Garonne River, looking for the inherited farm mentioned by Ausonius, Bordeaux's ancient poet, in *De Herediolo?* To sit at home, sipping our own Chardonnay with a pile of lobsters caught by a friend, or to pull a Roquefort tart from the oven while the aromas of an early Cabernet vintage met its pungency in the air, was justification enough for our labors. They might have been enough, too, to heal the damage done by the wear and tear of

229

work that never ended, and crises that were beyond our control, had it not been for something inevitable and heart-wrenching that happened: the death of Alex's mother.

Our children called Alex's mother "Granbetty," a name she had selected herself. It always bothered her that her given name was Betty—not Elizabeth, just Betty—a movie star's name. After she divorced Alex's father, a few years after we started the vineyard, Granbetty moved to the West Coast for a while. She returned when she realized how attached she was to her home town of Rochester. She was my size, about five-five and slender, with short, dark hair, a prominent nose, a short chin, and deep, deep hazel eyes. Granbetty was unique. I can see her now, pushing up the sleeves of her sweater so that her bangle bracelets clanged together, a mischievous smile on her face. She had the spunk of a teenager, an inordinately generous heart, and immense curiosity. At the same time, her insecurity made it hard for her to finish the projects that she started. She learned Braille and sign language, only to stop using them as soon as she mastered them. She took art and pottery lessons, and while her art was always of a commercial caliber, she was always changing mediums. Any time she became good at something, she'd abandon it. It seemed she was afraid to succeed. Always uncomfortable with herself, she tried to solve her disquietude through various therapies—psychotherapy, scream therapy, even starvation therapy. She actually paid a couple to keep her locked in their ranch house, out of touch with the world and with a minimal amount of food, for a few weeks. This deprivation therapy was supposed to be good for her psyche. She did have the sense to recognize the scam and flee. After she got the divorce that she had wanted, she deeply regretted having rejected her husband. He remarried, to someone who didn't complain that he was unresponsive or too conservative.

Granbetty visited us a couple of times a year. Anne and Zander loved her; she took them on excursions to the shore and made her own affection for them abundantly clear with bear hugs and laughter. She taught them the card games she had learned as a child during summers spent in the Finger Lakes—Russian Bank, Hearts, and Cad. I loved her, too, but it was difficult for me when she came to visit because she was so tense. She clanged her teeth and almost stood on my feet as she asked, "Lou, how do *you* like to cook lamb?" I knew she meant "I want to cook the lamb *my* way." Alex made himself scarce in his study much of the time that she visited—I imagine it was hard for him to deal with her because she asked a million questions about the details of the farm.

Granbetty was a smoker, and in 1986 she had a cancerous section of her lung removed, followed by radiation therapy. She made a good recovery, until the summer of 1992, when she started spitting up blood. Her doctor told her he'd watch her, but it was not until she collapsed in the airport on a Christmas visit to see Meg in Virginia that he really paid attention to her condition. By February 1993, she was losing weight, having more radiation therapy, and in severe pain. Alex, Charlie, Meg, and their oldest sister, Sue, agreed to take turns staying with her in Rochester while she was in treatment.

Alex and I arrived there in time to take her to her afternoon radiation treatment on a wintry Monday. She complained of pain in her neck. Her doctor listened to her and said, "I don't think the cancer has spread. Maybe you have some bursitis there. I'll give you cortisone for it." Hearing this, Alex and I asked him if he couldn't give her something for the pain. He refused. "I'm going away through the weekend," he said. "We'll do a CAT scan when I get back, and then we'll see."

The days that followed were as emotionally devastating as any I have lived. With Granbetty's lead doctor out of town, her radiologist also refused to give her painkillers, even though he admitted that he could feel a tumor growing in her neck. We convinced him to order a CAT scan immediately, but it had to be done on the other side of town, and we had to wait till Friday for it.

While we waited for the test, we tried to ease her pain. It was impossible. There was no way she could sit, stand, or lie down so that her tumors did not press on a nerve. Her bed was covered with antique pillows of all sizes, which I kept rearranging to prop up one part of her shrinking frame or another. In the daytime she made lists of things for us to buy at the grocery store even though we had already fully stocked her kitchen. She worked at knitting a sweater, but she would pull it apart and reknit it every day, like Penelope of *The Odyssey,* who tore apart her weaving every night in order to forestall her own fate. We took turns staying up with Granbetty through the night, as she cried for help that we could not give her.

When Friday came we drove through a blizzard to get her to the CAT scan. I held her in my arms so that she could endure the pain caused by the car's vibrations as we made our way in the snow. The test proved that there were tumors growing everywhere in her neck, chest, and shoulder. Armed with the CAT scan films, we finally got her radiologist to give her a morphine patch. At last, she slept for a few hours, and so did we.

Her doctor was embarrassed when he got home to find that his patient's family had taken charge of the case. From that point on, he gave her the medication that she needed. Granbetty was able to stay home, with hospice care, until she died in early May. Two days before she died, Alex called his friend Dr. Tom Cottrell for advice on her condition. Her own doctors didn't want to give us a prognosis for her, and Alex wanted to make sure that he visited her again before she died. Alex had planned to go to see her on the next weekend, but when he told Tom that his mother was semicomatose, Tom said, "Don't wait. Get your siblings together and go there now."

Meg, Charlie, Sue, and Alex were all by Granbetty's side when she took her last breath. Alex told me how he had held her hand and read from Percy Bysshe Shelley's "To a Sky-Lark" as she lay comatose before them. Quietly, he recited it:

> *"Waking or asleep,*
>   *Thou of death must deem*
> *Things more true and deep*
>   *Than we mortals dream,*
> *Or how could thy notes flow in such a crystal stream?*

> *"We look before and after,*
>   *And pine for what is not:*
> *Our sincerest laughter*
>   *With some pain is fraught;*
> *Our sweetest songs are those that tell of saddest thought.*

> *"Yet if we could scorn*
>   *Hate, and pride, and fear;*
> *If we were things born*
>   *Not to shed a tear,*
> *I know not how thy joy we ever should come near."*

When he finished reading, she opened her eyes, then closed them and was gone.

After Granbetty's memorial service, Alex and I stayed in a pew of the church and talked quietly with our children about her life. We wondered aloud why Granbetty had been so unhappy. She was strong and smart and beautiful. Why couldn't she be satisfied with her own talent, or her family's affection? What had happened to her that made her doubt herself? How did she keep her sense of humor even when she was in pain? There were many, many questions, and no answers.

In the months that followed her death I would sometimes find Alex at his desk, resting his head on his long arms, with John Prine playing in the background.

The life of the vineyard continued largely as before; only

now I began to take on some of Alex's community obligations. When the supervisor of Suffolk County invited him to serve on the Long Island Regional Planning Board in 1994, Alex turned the job over to me. In addition, I took on his role in the Wine Council, planning events and seminars. When all but the most prominent visitors came, I took care of them. I went on trips with the kids—even vacations to Florida with his father and siblings—without Alex. He didn't like to travel—the beds were never long or firm enough, the airplanes' pressurized cabins hurt his ears, he didn't like to follow an itinerary—and I did. By now he was so deeply involved in his studies and his private thoughts that I left him alone. We still worked well together when it came to making wine, but our disagreements about how to tend the vineyard became a source of conflict. I got up early every morning to exercise and then to meet the crew by eight A.M. At that hour Alex was of no mind to hear my requests for extra help, equipment repairs, or more timely delivery of supplies.

Doing so much of the physical work of the vineyard myself made me feel my own strength more and more. In the beginning I had always relied on Alex to advise me on everything—what to say on a tour, how to describe a wine, how to adjust the labeler, how many buds to leave on a spur. Now I just figured it out for myself. I stopped jogging when my knees gave out but I took up aerobics, then step training, and finally yoga. Without my noticing it, my sense of self-reliance moved from my muscles to something more fundamental. When my yoga teacher described Kundalini, the serpent, rising inside the body as an internal force, one that would link me to a greater power than anything I could see, I thought again of Dylan Thomas, and the "force that through the green fuse drives the flower." Still, I didn't know what my growth connected me *to*. With all this power, why did I feel so alone?

Not long after Granbetty's death, I made a serious mistake in winemaking. Although it was an innocent error, it made me feel as if I couldn't rely on myself after all.

We had harvested fruit of especially high quality, so we chose to use a particular yeast that had been developed by a reclusive scientist who would release only small amounts of it at a time. We had used it once before and loved the spicy quality it gave our Merlot and Cabernet Franc. Most wine yeasts start very vigorously, but this one is lazy, and it takes a scary amount of time to get going. When we didn't see the reduction in sugar that we normally expected in our fermentation, we decided to add some nitrogen to the tank, a normal way to give yeast a boost. I went down to the lab to measure out the diammonium phosphate (DAP) that we used as a source of nitrogen. I had ordered the DAP from the same source every year; it came in the usual cardboard box with a plastic bag that held the white crystals. The box was clearly labeled "diammonium phosphate" by the company that sold it to us.

I actually enjoyed weighing out the things we used to improve our wines. Besides yeast and yeast nutrients, there were pectic enzymes that made the juice flow freely; bentonite, carbon, gelatin, and egg whites for clarification; and potassium metabisulfite, which yields SO_2 (sulfur dioxide), a preservative. The only one I didn't like to handle was the SO_2. Like sugar or DAP, it's a fine white crystal; unlike them, it has a chokingly sharp pungency. Of all food preservatives that have ever been tried, it is the only one that will protect food from both oxidation and bacterial spoilage. The drawback is that it interferes with thiamine in the human respiratory cycle. In people who are sensitive to it, including myself, SO_2 will produce a headache. To avoid this, and also to protect myself from inhaling the dust and vapors of all the things I measured, I always wore a gas mask while weighing them out.

When wine ferments, it creates SO_2 on its own. Whatever SO_2 a winemaker adds is a supplement to something that is naturally occurring in the wine. The problem is that it is virtually impossible to make a wine that will last with any quality for more than a few months without adding SO_2. Wines made all over the world have had SO_2 added for centuries. Before granular metabisulfite was used, winemakers would just burn a sulfur wick inside a barrel and then add wine to the gas. When Congress legislated warnings on cigarette packages, the senators from the tobacco states got a compensating law passed that required wine producers to label the presence of sulfites in wine. Consumers were alarmed by this; few were aware that at the same time the wine industry had vastly reduced the use of SO_2 by using better sanitary precautions in winemaking. Golden raisins and orange apricots, as well as foods like fig cookies, have far greater doses of sulfites in them than most wines.

We tried to minimize our use of SO_2 in every possible way, and because we didn't have to transport our fruit from distant vineyards or deal with high-temperature storage conditions, we learned that we could eliminate SO_2 at the crusher entirely. Our wines had remarkable natural stability, but we still had to add a minimal amount—maybe 40 parts per million—to every batch that was in storage. As much as I disliked it, I was very familiar with it. Although I wasn't as concerned about handling DAP as about handling SO_2, I still put on my mask when I went to measure the DAP for the stuck wine. I took it in batches for each tank to Alex, who carefully added it while stirring the must. The next day the wine still wasn't fermenting, and the following day was no better. We decided to add more DAP.

This time, after Alex had added the DAP, he came looking for me. "Are you sure that was DAP you measured out?" he asked. "It smells like SO_2 to me."

"Of course I'm sure," I said. "The box is clearly labeled."

Together we went back into the lab and opened the box. Without a mask on this time, I could smell the unmistakable odor of SO_2. The purveyor that had sent us the box labeled DAP had erroneously put SO_2 in it. With every addition of what we thought was DAP, instead of encouraging the yeast to grow, we were killing it. Not only that, but the rate for DAP is far greater than the rate for SO_2, so now we had almost the legal limit of sulfites in our wine. I had made an honest mistake, but it was also a careless one. The only way we could use those wines was to blend them with other unsulfited wines or juice. What had been our top wine of the vintage ended up being a wine that we sold in bulk.

Back in the vineyard, where I always felt happy and comfortable, I was working with my crew of three or four women when a small helicopter came flying directly over our heads. It was not uncommon to see these small copters—they were either crop dusters for the potato farmers, military surveillance for the airbase in nearby Westhampton, or real estate agents scoping us out. One of my team called out, "There's something wrong up there. It's making too much noise." We all looked up and saw the aircraft burst into flames. As it dropped from the sky billowing black smoke, one crew member ran to the house to call 911, while the rest of us ran to the field where it came down.

Somehow the helicopter landed upright. The flames subsided, and the pilot climbed out. Rushing to see if he was all right, I shouted to reassure him, "Help is coming!" The sirens of fire trucks and police cars were already wailing from the direction of the village.

"No! No!" the pilot cried out. "Why did you call them? I don't need any help."

The man was limping and covered with soot, but he was livid at the thought that anyone would find out he had crashed. "Why are you angry at us?" I asked him. "Would you be angry if you had really been hurt? How were we supposed to know that your flaming helicopter would land safely?"

I never found out what his problem was. Maybe he was running drugs or something. When I think about it now, I can see that Alex and I were in the same situation as that pilot: our vehicle, our marriage, was in danger of crashing, and the last thing we wanted was for anyone to know it. I didn't know it myself. But Alex did.

15

Moving On

WHEN I HAD READ THE LITTLE HOUSE BOOKS AS A CHILD, I had wondered why Laura Ingalls Wilder's father, Charles, had moved his family so many times. It seemed that every time they got settled in a place—got the garden going, the livestock breeding, the house finished—Charles would want to move. I thought they should have stayed put. There must be some restlessness intrinsic to being a pioneer. And I felt that Alex was restless. In the case of our vineyard, maybe it was in the nature of the place that the same problems came up again and again—those weeds, those extra expenses, those accidents. Since my outlook had always been that if I tried hard enough, I could eventually get what I wanted, I thought that we could work around those problems. My nose was very, very close to the grindstone.

Alex became more and more involved in researching theories about the origins of language. He went back to his old interest in Chinese and other languages. It could have been the influence of the evolutionary theorists that Alex read—the survival of the fittest being linked to the fight-or-flight response—or maybe it was a matter of getting past his fiftieth birthday, or a flashback to feeling confined in his body cast, or something that I will never know that caused Alex to do what he did next. Certainly, I sensed that there was a growing distance between us, but we had always had periods of unrest with each other that usually cycled with the seasons. I thought I could count on soothing words and a soft touch to repair whatever hurt had caused our distance. That had worked for almost thirty years.

In the winter of 1997 I suggested that Alex rent a house in New Hampshire that was owned by an old friend of ours so that he could continue his research projects completely undisturbed instead of being constantly interrupted by the demands of our business. There wasn't much going on in the vineyard in the depths of winter, and I would hold down the fort at home.

It was early in the day, a couple of days after Alex had gone to New England, when Betty rushed into the house to find me. "Call 911," she shouted. "Mark fell off a ladder in the winery."

I made the call, then rushed down to the cellar where Mark, our cellar master, was lying stock-still on the concrete floor. The ladder that had been propped against a tank was lying next to him. "Mark, are you okay? Don't move!" I said to him, after I saw that his eyes were open. "Can you feel your toes?" I asked him.

Fortunately, he could move his toes, but after the rescue squad took him to the hospital, we learned that he had broken his back. It was a clean break, and although his spinal nerves were bruised they were not torn. Still, he would be sidelined for months.

Although I had promised Alex that I wouldn't call him in New Hampshire—the whole point of his being there was to be left undisturbed—I picked up the phone now and dialed the number there. The phone rang and rang and rang. It didn't surprise me that he didn't answer it. Still, I thought that if I tried it again, he would know that someone was really trying to reach him and maybe he would answer it. It rang and rang again. I was just about to hang up when a voice I didn't recognize said hello. It was a carpenter who was there to do some work on the bathroom. When I asked to speak to Alex, the man said, "No one is here. There is snow on the ground and no footprints. No one has been here but me."

It was two days before Alex checked in with me and came home. It didn't matter where he had really gone. What mattered was what he said to me: "I want a divorce."

A divorce. A divorce. I doubled over, hyperventilating. Suddenly I was standing at the bottom of a deep hole with voices above me shouting, "You're not good enough!" That's how I felt. And then I was suffocating in my own tears. I pictured shovelfuls of soil being dumped on top of me. I pictured the worms in the soil blithely ready to do their work. I think I crawled away to bed—but what bed? Where to sleep? I don't remember where I went, but the next day I started playing a recording in my head that was my version of a Muppets' *Musicians of Bremen* record that my children had loved. In this version, Kermit, Miss Piggy, and the other Muppets play the old farm animals whom nobody wants. Kermit keeps saying, "I'm beat up, worn away, throooowed out" in his nasal frog voice.

For the next few months life went on as if nothing had happened except that Alex didn't touch me and I moved up to Zander's bedroom, which he had vacated while in boarding school. As if I had been bitten by a snake that releases its poison slowly, I couldn't believe that the bite was mortal, that my marriage was already dead. I went to the library and read a little divorce law, enough to tell me that New York State does not permit divorce on the basis of irreconcilable differences. One spouse must have substantial grounds to divorce the other. In our case, the only option was "constructive abandonment"—no intimate relations for a year. We didn't discuss it at all, and I just numbed myself to life for the duration. There was nothing to discuss really. And besides, I felt as if all the blood had been sucked out of me. It was all I could do to continue going through the motions of work, day by day. I knew our days in the vineyard were numbered. All my equity was tied up in the vineyard, so I wouldn't be able to buy Alex's share when the year was up.

Then I had an idea. When I was fifteen I had gone with my family to visit friends in Montana. There the night sky had a weight that made me feel as if I could touch the heavens. In the early years of the vineyard, when we had slept outside all summer long, there had been a few nights when I was reminded of that feeling, but most of the time there was a haze of humidity obscuring the stars. I made up my mind that I had to go back to Montana. I needed to be in the wilderness again. That's how I would celebrate my fiftieth birthday. I knew it sounded trite—going to the wilderness to solve a midlife crisis—but at that point I felt so isolated all I could think of was that I had to get away. I would seek solace where I had found it before, under the stars.

As shattered as I was, I didn't want to have to plan the trip myself, so I signed up with Outward Bound for ten days of backpacking in the Absaroka wilderness. Outward Bound is a survival training course that had initially been designed to teach young men self-reliance—perfect for me!

My group had seven women, three men, and two female guides, almost all of them in their thirties. It was a very fit group; several of them had run marathons. We had been sent instructions of what clothes to bring, but even so we had to spread everything on the ground and eliminate most of it before filling our packs. It wasn't worth carrying the weight of a change of clothes. As it was, our packs weighed over sixty pounds. The weather was hot during the day but almost freezing at night. We carried sleeping bags, tents, ten days' worth of food for twelve people, ropes, rain gear, water bottles, and helmets to wear in case of rock slides.

Our guides taught us how to put moleskin on our feet to

stave off the blisters that would inevitably hobble more than half of us. They taught us how to tie our food high in the tree branches, away from bears at night, and how to look bigger in case we met a bear in daylight. We learned to crouch on top of our packs during the lightning strikes that came nearly every afternoon. And every day, we all had to know who was carrying the snake-bite kit.

I wasn't afraid of the bears, or the lightning, or the snakes. I listened attentively to all the safety instructions. Our guides were ultracautious. My trouble began when I strapped on my backpack. I weighed less than 115 pounds, and climbing to an altitude of eleven thousand feet made the pack feel even heavier. The packs were quality gear, but the smallest size available was a medium, which I couldn't tighten enough for the weight to rest on my hips as it should have.

Somehow I managed to climb the first part of the trail, which went almost vertically to a high alpine ridge. I thought the top would be easier, but the ground was covered with fist-sized rocks that shifted as I stepped on them. Tears streamed down my face, but I was determined not to quit, so I kept silent. Soon I was staggering far behind the others.

One of the women in the group was a wiry Brazilian named Angelica. She didn't weigh much more than I did, but she swooped next to me and offered to take part of my burden. I gave her half the tent poles I was carrying, and even though they weighed only a couple of pounds, this made all the difference. One of the guides, Kate, taught me how to coordinate my breath with my steps: inhale as one leg goes forward; exhale with the effort of transferring weight to that foot and pushing ahead. Like planting vines with a pregnant belly, I realized. Kate watched me totter on the rocks and said, "Look ahead two steps, not one. If you lose your balance on the first step, you will already know where to regain it on the next."

That was a revelation to me. Just by shifting my gaze slightly ahead, I could stride forward with confidence. It wasn't a physical limitation that held me back, but a mental one. No matter how strong a prescription I had in my glasses, I would never have been able to see the way I could now. It was all a matter of balance.

Another part of the wilderness balancing act was to stay hydrated. It was so dry up there that we had to drink water constantly. When we got past the tree line there was no place to hide to relieve ourselves, so we got used to peeing behind boulders. More serious business had to be buried, and there was great demand for the group's only spade every morning.

Burying human excrement was vitally important to our "leave no trace" camping ethic. One morning when someone went looking for a place to poop and virtually tripped over a heap of turds that had been left by a previous camper, we were all called to observe, and then to bury it. We stood around it like a solemn prayer group, bemoaning other people's negligence and promising God that we would never leave our shit unburied. I felt like singing a hymn, but I couldn't think of the right one. Righteously willing to cover the sins of our brethren, we carefully dug a hole and laid the turds to rest.

In the evening, after a full day of hiking, as we lay exhausted under the stars, Kate asked, "How many of you are willing to take risks?"

I am, I thought. The vineyard had been a risk. I had gone forward with it, all burners firing. I had been the pioneer mother, giving birth in a van. I stared at the night sky with a nagging feeling that I was missing something. I wanted to think of myself as a risk-taker, but what had I really been? There hadn't been one thing I could think of that I had risked without Alex. Wasn't he the brilliant one, who could make the important decisions? I would back him up one hundred percent, but without him, wouldn't I have had a more ordinary life?

Angelica and I started to talk about ourselves. I couldn't admit to her that I was headed for divorce. It was too new; too raw to admit. I hadn't even gotten up the courage to tell my mother yet. When Angelica told me that she felt like the black sheep of her family, I told her I felt like I wasn't good enough. I said that I felt an anger inside me that didn't belong to me. It felt like something I had inherited, and I couldn't make it go away. "I've been angry, too," Angelica said. "The truth is, when we look at other people, we are looking into a mirror at ourselves." When I heard that, I realized that I had been unable to look Alex in the face since the day he had told me he wanted a divorce. Maybe I was afraid of seeing myself there.

The next day, when I found myself lagging again at the end of the trail, I decided that after our next stop for water I would try taking the lead. It was high time I took charge of myself. I didn't ask permission; I just put on my pack and started up the trail first in line. The guy behind me was one of the marathoners, but he didn't complain about slowing his pace. In fact, my pace increased as the trail opened before me. I noticed that others who had been faster than I had been when they were in the lead now started to lag behind.

We spent the last night alone, each of us on a separate knoll or grove of twisted trees. We were out of sight of each other, high in the alpine meadow that was defined by bordering boulders and snowmelt streams. Like the hermits of a Renaissance painting, we could have called out to each other, but instead we sat silently, awaiting God's call.

When God didn't call right away I decided to pass the time drawing. I took out my notebook and pencil, sat on a rock, and tried to make a botanical drawing of a broad-leafed plant that had a small yellow flower at the end of a spindly stem. It occurred to me that this plant was a cousin to one of the weeds that I had smitten with my hoe every August in the vineyard. Now I could look at it without judgment—a shape to copy on paper, a life force of its own.

Before the afternoon thunderstorm came I needed to rig my tarp as a shelter. The tarp was a rectangle of blue plastic with grommets at the corners. It was supposed to have nylon ties to attach it to a tree or shrub, but my tarp was missing all but two ties. Remembering that in the vineyard if I needed a nail, all I had to do was to look in the driveway and one would emerge, I looked at the ground and saw my boot laces. There were laces in the sneakers I carried, too. Removing all four laces, I threaded them through the tarp grommets and rigged my pup tent. Then I braided chains of those flower stems to make new laces for my sneakers. What ingenuity! With those skills what couldn't I do?

After dark I crawled into my sleeping bag under the tarp. It didn't take long for me to fall asleep, but I was awakened by the sound of something scratching outside. A bear! My body stiffened. I knew he could smell me. Was he hungry? Was he hostile? The scratching sound stopped and I went back to sleep. A while later I awoke and it started again. And again I stiffened. As I lay there, trying to figure out what to do about this bear, I relaxed my foot. There was the scratching sound again. Feeling my toe press against the tarp, I realized that there was no bear. There was only the sound of my feet rubbing the end of the tarp.

Dawn came and I sat cross-legged on the grass, looking at the mountains and the sky. A great peace fell over me. Nothing could ever be better than this, I realized. Maybe I was a person who was easily fooled. Maybe I relied too much on others. Was I arrogant to think that I could solve all my problems, just by trying harder? At that moment, as the sun rose, I laughed.

Epilogue

THE YEARS BETWEEN THE BEGINNING OF OUR DIVORCE AND
its resolution in October 1999 are years that I would just as soon
forget.

Part of the shock that came with the word *divorce* was my real-
ization that when our marriage was over, my life as a vintner
would be over, too. Alex had other projects that he wanted to
pursue, but I didn't know what I wanted. Maybe I could have
found an investment partner to help me keep the vineyard going, but that would have entailed putting all I owned at risk
with someone I didn't know. Besides, the vineyard wasn't just a
business. Our home and family were all wrapped up in it. When
Anne was little, she tried to say, "We have each other"; only it
came out "We have the jother." It was "the jother" that had kept
us going, and now we didn't have it.

Alex and I moved to separate ends of the house until the vineyard was sold. He would go away for a few days, and when he
came back, I would go to visit friends or family in Boston, New
York, or Washington. It was a revelation that I could leave for
three days at a time and the vineyard would still be functioning
when I returned. Anne and Zander returned from other jobs and
college to the vineyard for the final months, interrupting the

normal course of their lives to help us. I told them that I felt it was an unfair burden for them to be there, and they didn't have to do it, but they were adamant that it was their unfinished business, too. They wanted to bring in our last harvest.

For a time, I was engaged in a mental spacewalk, as in David Bowie's ballad beginning "Ground Control to Major Tom," but my connection to my friends at "ground control" was better than Major Tom's. Once I learned to breathe the rarefied atmosphere of freedom, I could understand Alex's urge to be free and let him go. We continued to make wine together with as much goodwill, if not as much delight, as we ever had.

When we listed Hargrave Vineyard for sale, we were instantly besieged by members of the press who reported that we were moving on. They didn't know we were getting divorced, but they made it sound like we had already packed our bags. The local paper featured a cartoon that played upon our references to the ancient Romans. As a parody of Julius Caesar announcing his conquests with *"Veni, vidi, vici"* ("I came, I saw, I conquered"), it showed the two of us driving off into the sunset together, with the caption *"Veni, vidi, vino!"* It was meant to be triumphant, but I knew the truth; and if the truth wasn't quite defeat, it sure felt like it.

Listing the vineyard for sale was a far cry from actually selling it. We tried to screen potential buyers, but there was still a cavalcade of dreamers and opportunists who came around to check out the property. Among these were a multimillionaire who was relocating from Paris and wanted a pied-à-terre, a California investment capitalist who wanted to siphon off excess California wine into our market, and a man with limited knowledge of English who wanted to establish a fish farm and had somehow confused "grapes" with "fish." Another prospective buyer came from New Jersey by limousine to inspect the property with his obese teenage son, two aides, and an armed bodyguard. He kept smacking his son on the head every time he wanted to show him something. The lad did not appear to be impressed with our establishment.

We had several offers for our property, which fell apart when negotiations got mired in complex details. But one fine day an Italian prince named Marco Borghese appeared on the scene with his stunning wife, Anne Marie, and their two children, Allegra and Giovanni. At last, here was a real Prince Charming. When Marco let his kids ride through the vines on the front of his car, their joyous laughter filled the vineyard, and I knew it could be a home for them as it had been for us.

On October 22, 1999, the Borgheses bought Hargrave Vineyard, including our house and all our wine. Our divorce had been finalized just a few weeks before that. By then, I had sorted every item that we had shared during thirty-one years of marriage. Off in the far corner of one of our barns there was a trove of wedding presents we had never used. A silver-plated *cendrier*, intended to be used by the butler in emptying ashtrays from the dining room table, just never came in handy. Two sets of highball glasses had been redundant, and useless to us since we quit drinking hard liquor in 1969. Then there were toys too dingy to offer a new baby, and books that had been dreary to read even when we had been assigned them at college.

What we didn't give away or throw away, we offered at a yard sale, where everything was carefully displayed in the back of the visitors' center. Zander fled from the scene to visit his friends upstate. Anne helped me make change for customers, controlling her dismay as people pawed over the relics of her childhood.

By the time we closed our sale to the Borgheses, Alex had found a place to live off Long Island that suited him, with hills and a lake, but I didn't know where to go. Part of me wanted to flee, but where to? I loved Long Island.

It wouldn't be fair to leave Anne and Zander in the lurch after selling the vineyard, so I promised them that I would find a place where they could stay, too. The problem was that there were so few places to rent on the East End that my choices were limited. I couldn't afford to commit to anything until our sale closed, and the closing date was moved several times. When it finally came, I had seven days to vacate the vineyard with my remaining belongings. At the last minute, I was able to rent an old farmhouse nearby. The wind whistled through its cracked windows, but it was a roomy way station until I could find a house of my own a year later.

The new millennium came, and I went on an extended trip around the world. I wanted to visit vineyards on the other side of the planet, in Australia and New Zealand. Vineyards, wherever they were, seemed familiar to me, so no matter how far I was from home, I felt connected. Along the way, I realized that I still didn't know how to be alone. But before I left, I had made a vow to myself that I would not to wait for someone else to lead me. There were things out there to see and do, and I would just have to set one foot ahead of the other, look two steps ahead, and go.

Now I am back on the North Fork, enjoying my life surrounded by friends and grapevines. From time to time, Alex and I get together for a cup of tea or a meal with our children. We talk about the old days in the vineyard, but we don't discuss what happened between us. I know we were lucky for all that we had and all that we accomplished. We began our venture with the hope and naïveté of all pioneers, and we really did cross the prairie before our Conestoga wagon broke down.

When I think back on it, the closeness that developed between Alex and me as we brought our vineyard to fruition was not always compatible with the restlessness of our pioneering spirits. What had felt secure became claustrophobic. In its complexity, our marriage was not unlike a great wine, always having certain elements both agreeable and disagreeable, appearing in a state of flux. This same sort of changeability, untenable in marriage, is something I still look for in wine, as I crave the sensory stimulus of complex smells and sights. At home now, when I open a bottle of the wine we made together, scents of cinnamon, coffee, olives, currants, cedar, and cigars all emerge. As I swirl it in my glass, hints of other things emerge, as if smelling could unfold history, placing me on the dock in Venice when Marco Polo returned from China. Silk, velvet, corduroy, and cotton are all there in the texture of wine, and the color of the wine mimics hues of rubies, topazes, and blood.

I have a little garden now, next to a peaceful house with an expanse of farmland all around it. I don't grow much, just flowers, herbs, and a few tomato plants. When weeds come up, I laugh. It's nice to have just a small plot of weeds.

Notes

CHAPTER 1. THE HOMESTEAD

18 **"I have tried"** Ulysses P. Hedrick, *The Grapes of New York,* New York State Department of Agriculture, 15th Annual Report, 1907 (Albany, N.Y.: J. B. Lyon, 1908), vol. 3, part 2:6, p. 22.

CHAPTER 2. FARM LABOR

33 **"Yesta polska"** Properly spelled *Jeszcze Polska nie zginela, poki my zyjemy.*

CHAPTER 3. RESTLESS BEGINNINGS

65 **"under his vine"** Micah 4:4.

CHAPTER 4. WESTWARD HO! AND BACK

78 **"Good wines"** William Massee, *McCall's Guide to Wines of the Americas* (New York: McCall Publishing Co., 1970), p. 4.

78 **"Now balance with these Gifts"** John Dryden, trans., *The Works of Virgil: Containing His Pastorals, Georgics, and Aeneis* (London: Jacob Tonson, 1697), p. 71.

80 **"Moist earth produces"** Ibid., p. 81.

81 **"They Have My Name"** Quoted in Taylor's obituary, Frank J. Prial, *New York Times,* May 2, 2002.

CHAPTER 9. GROWING UP

154 "The force that through the green fuse" Daniel Jones, ed., *The Poems of Dylan Thomas* (New York: New Directions, 1971), p. 73.

CHAPTER 11. THE WORLD AT OUR DOOR

178 "Because they are so young" Frank Prial, *New York Times Magazine,* July 9, 1978.

184 "The countess lifted her tulip-shaped glass" Dallas Gatewood, "Praise for an LI Wine," *Newsday,* Nov. 20, 1981, p. 7.

CHAPTER 12. ARROWS IN THE BACK

196 "I wish to comment" This and all subsequent quotations of Tomkins are from a letter from John Tomkins to Louisa and Alex Hargrave, October 6, 1997.

FOR THE BEST IN PAPERBACKS, LOOK FOR THE Ⓟ

In every corner of the world, on every subject under the sun, Penguin represents quality and variety—the very best in publishing today.

For complete information about books available from Penguin—including Penguin Classics, Penguin Compass, and Puffins—and how to order them, write to us at the appropriate address below. Please note that for copyright reasons the selection of books varies from country to country.

In the United States: Please write to *Penguin Group (USA), P.O. Box 12289 Dept. B, Newark, New Jersey 07101-5289* or call 1-800-788-6262.

In the United Kingdom: Please write to *Dept. EP, Penguin Books Ltd, Bath Road, Harmondsworth, West Drayton, Middlesex UB7 0DA.*

In Canada: Please write to *Penguin Books Canada Ltd, 10 Alcorn Avenue, Suite 300, Toronto, Ontario M4V 3B2.*

In Australia: Please write to *Penguin Books Australia Ltd, P.O. Box 257, Ringwood, Victoria 3134.*

In New Zealand: Please write to *Penguin Books (NZ) Ltd, Private Bag 102902, North Shore Mail Centre, Auckland 10.*

In India: Please write to *Penguin Books India Pvt Ltd, 11 Panchsheel Shopping Centre, Panchsheel Park, New Delhi 110 017.*

In the Netherlands: Please write to *Penguin Books Netherlands bv, Postbus 3507, NL-1001 AH Amsterdam.*

In Germany: Please write to *Penguin Books Deutschland GmbH, Metzlerstrasse 26, 60594 Frankfurt am Main.*

In Spain: Please write to *Penguin Books S. A., Bravo Murillo 19, 1° B, 28015 Madrid.*

In Italy: Please write to *Penguin Italia s.r.l., Via Benedetto Croce 2, 20094 Corsico, Milano.*

In France: Please write to *Penguin France, Le Carré Wilson, 62 rue Benjamin Baillaud, 31500 Toulouse.*

In Japan: Please write to *Penguin Books Japan Ltd, Kaneko Building, 2-3-25 Koraku, Bunkyo-Ku, Tokyo 112.*

In South Africa: Please write to *Penguin Books South Africa (Pty) Ltd, Private Bag X14, Parkview, 2122 Johannesburg.*